Preface

Biometric technologies including iris, voice, fingerprint and vein pattern recognition – once the realm of science fiction and film – are now becoming more of a daily reality. The potential biometrics provides as an individual identifier has resulted in the widespread diffusion of this technology into people's lives. For example, Irish citizens now encounter biometric applications in many different situations, such as for workplace time and attendance, for physical and logical access (*e.g.* for laptops) and particularly in relation to international travel. In addition, Ireland is now recognised as a European hub for biometric research in both industry and academia. Indeed, the European Biometrics Forum (a European-wide stakeholder group supporting the appropriate use of biometric technologies) was established in Dublin and is part funded by the Irish government.

Given Ireland's involvement in and contribution to biometric research, the Irish Council for Bioethics (the Council) considered it appropriate to examine the ethical, social and legal issues associated with biometric technologies and the collection, use and storage of biometric information. This opinion document outlines the Council's views and recommendations on these issues. In contrast to many emerging technologies, biometrics has the potential to impact peoples' lives directly at the level of the individual, but also from a larger societal perspective. Therefore, the Council considers it imperative for there to be increased trust, transparency and honest engagement with the public in relation to biometric technologies and applications. The Council is hopeful that this document can provide a useful and informative resource for both policy makers and the general public and will also help to engender greater discussion and consideration of the issues pertaining to biometrics.

I would like to express my thanks to the members of the Rapporteur Group and the Council, as well as to the Council's Secretariat, for the time and effort they have expended in the production of this document. The Council would also like to thank both the stakeholders and the members of the focus groups who contributed to the consultation process. Their input proved very helpful in the Council's deliberations on this rapidly developing and multifaceted topic.

Dolores Dooley

Dr Dolores Dooley
Chairperson
Irish Council for Bioethics

Foreword

This opinion document examines, in detail, existing and forthcoming biometric technologies, and the ethical and legal ramifications that are relevant to this technology.

Developments in biometrics hold very great hopes of ensuring the protection and maintenance of one's identity and privacy. Biometric technologies also have the great potential to enhance security and safety of citizens.

However, along with this potential, is also the potential to compromise the privacy of individuals in the misapplication of the technology. Central to this discussion are the themes of balance, proportionality and transparency. The Council recognises benefits of biometric technology however in alliance with these themes.

Any technology which purports to collect and retain personal information in relation to an individual should be used only after careful consideration and only as to its necessity as is required for the intended use. In order for individuals to have confidence that this technology will properly provide the positive benefits which it can, manufacturers, policy makers and end users of this technology should be as transparent as possible in relation to the technology, the reasons for its use and how the information obtained from the technology is to be secured. The autonomy of individuals again features in this opinion document of the Council: in relation to biometric technology, the autonomy of the individual should be protected by allowing an individual's involvement in the use of the technology, as much as is possible. There may be circumstances that are exceptions, where in the common good/public interest, the individual's involvement may have to be limited. However, these circumstances should be limited and properly considered and discussed before the invocation of such exceptions.

As is the Council's normal practice, it engaged in a consultation process by way of focus groups. The results of this exercise are to be found in the appendices of this report. In addition, the Irish Council for Bioethics organised a conference in November 2008 entitled 'Biometrics: Enhancing Security or Invading Privacy?'. The Council is very grateful to the participants at the conference. The views of our distinguished speakers have assisted as part of the process of the preparation and consideration of this opinion document.

The Council is also grateful to all those who made submissions to it (a list of submissions received by the Council can be found in the appendices).

It is the Council's hope that this detailed document will provide information, insight, and guidance to the public at large and to any party that is likely to be involved in biometric technology. We also hope that the document will act as a catalyst to further discussion of this topic at both national and international level.

The Council's terms of reference are to identify and interpret the ethical questions raised by biomedicine in order to respond to, and anticipate, questions of substantive concern and also to investigate and report on such questions in the interest of promoting public understanding, informed discussion and education. The Council hopes that this document achieves such ends.

The Rapporteur Group and the Council would again like to extend its gratitude and appreciation to the members of the secretariat, Dr Siobhan O'Sullivan, Ms Emily de Grae, Mr Paul Ivory and Ms Emma Clancy. Their efforts and diligence are invaluable to the Council's work past, present and future and to the completion and compilation of this opinion document.

Silence and inaction have very rarely served, if ever, to promote progress and prosperity in any given field. The Council hopes that it can continue fulfilling its objectives through its terms of reference in order to play its part in ensuring that constructive debate and activity in the field of bioethics continues and prospers. We hope that this opinion document is another small, but helpful step in that process.

Professor Alan Donnelly

Mr Stephen McMahon

Mr Asim A. Sheikh, BL (Vice Chair)

Rapporteur Group on Biometrics
Members, Irish Council for Bioethics

Table of Contents

Preface i
Foreword ii
Executive Summary vi
Chapter 1: An Overview of Biometrics and its Applications 1
What are Biometrics? 2
Why are Biometrics Used? 3
Architecture and Design of Biometric Recognition Systems 4
Practical Considerations 8
Technology Associated with Biometric Systems 14
Chapter 2: Overview and Comparison of Biometric Modalities 17
Market Overview 18
Fingerprint Recognition 19
Palm Print Recognition 25
Hand Geometry 26
Vein Pattern Recognition 29
Facial Recognition 31
Facial Thermography 35
Ear Geometry Biometrics 36
Iris Recognition 37
Retina 41
Gait 42
Odour Recognition 44
Voice Recognition 44
Keystroke Dynamics 48
Dynamic Signature 49
DNA (Deoxyribonucleic Acid) 50
Multimodal Biometrics 54
Future Biometric Modalities 58

Chapter 3: Ethical Considerations for Biometric Information and its Associated Technologies 59
Privacy 61
Personal Privacy 64
Informational Privacy 71
Autonomy 82
The Common Good 87
Proportionality 90
Trust and Transparency 94
Chapter 4: Biometrics Legislation/Regulation 99
Europe 103
Ireland 107
United States of America 109
Australia 112
Canada 114
Appendix A: Report from the Focus Groups 118
Appendix B: Biometrics Information Leaflet 132
Appendix C: Submissions Sought by the Irish Council for Bioethics 138
Appendix D: Submissions Received by the Irish Council for Bioethics 139
Rapporteur Group on Biometrics 140
The Irish Council for Bioethics 141
Abbreviations 142
Legal Instruments and Regulations 144
Glossary 147
Bibliography 151

Executive Summary

Recent years have been characterised by a more stringent requirement for people to be identifiable in response to security threats and to combat the escalating problems of identity theft. This increasing need to determine who an individual is has resulted in substantial growth in the implementation and use of biometric applications. Biometrics, as physical or physiological features, or behavioural traits, represent "something you are" rather than something you have (*e.g.* an identity [ID] card) or something you know (*e.g.* a password or personal identification number [PIN]) and as such are considered to provide a more robust confirmation of a person's identity. While people are, generally, more familiar with certain biometric modalities such as fingerprint, face, iris and signature, numerous other modalities are also being used, including voice, hand geometry, odour and gait (manner of walking).

Prior to being utilised for biometric recognition, a given physiological or behavioural characteristic is usually evaluated against the seven pillars of biometrics, namely: *universality* (all individuals should have the characteristic); *distinctiveness* (ability to distinguish between different individuals); *permanence* (should remain largely unchanged throughout the individual's life); *collectability* (relatively easy to be presented and measured quantitatively); *performance* (level of accuracy and speed of recognition); *acceptability* (an individual's willingness to accept the particular biometric); and *resistance to circumvention* (degree of difficulty required to defeat/bypass the system). Based on such evaluations it is clear that there are strong and weak biometric modalities, with the stronger biometrics meeting more of the seven criteria.

Biometric modalities offer a stronger assurance of identity as they cannot be lost or forgotten, they are difficult to copy, forge or share, and they require the individual to be present at the time of identification. However, biometric systems are not infallible and they are prone to errors and are vulnerable to attack. Since biometric information is an integral part of an individual, the potential to misuse or abuse this information poses a serious threat to privacy. Depending on the practical and technical measures taken during the design, implementation and operation of biometric systems, concerns relating to privacy may be diminished.

Considering the developments in biometric technologies, the increasing incidences of their deployment, and the diversity of their applications, the Council considers it imperative that the ethical, social and legal issues pertaining to the use of biometrics are examined and discussed. Similarly to other developments in science and technology, the challenges posed are not with biometric technologies *per se*, but in the manner they are applied and how the resulting data are dealt with. The use of biometric systems and applications raises a number of ethical questions, particularly issues of human dignity and identity (individuality) and basic rights such as privacy, autonomy, bodily integrity, confidentiality, equity and, in the case of criminal investigation, due process.

The Irish Council for Bioethics is of the view that, when implemented appropriately and managed correctly, biometric technologies can both improve security and enhance privacy.

However, this positive view of biometrics is tempered by the knowledge that these technologies could have significant implications for an individual's privacy. Consequently, the Council places paramount importance on respecting and protecting an individual's autonomy as well as his/her personal and informational privacy with regard to the collection, use and storage of his/her biometric and other personal information.

In particular circumstances, the Council acknowledges that it may be appropriate to override certain individual rights for the benefit of the common good. However, the Council is concerned that this principle may be over utilised in order to implement certain applications without adequate justification. To be justified, the Council takes the view that a biometric application must represent a proportional response to meeting the challenge at hand. This requires providing a detailed rationale for the necessity of using biometrics as opposed to some alternative technology or methodology. Given the concerns raised in relation to biometric technologies, the acceptance of, and trust in, such technologies and those operating them requires transparency and accountability in conjunction with dialogue and feedback between all the parties involved.

The Council considers it imperative to determine whether or not a biometric application is essential and can be justified, prior to its introduction. Each application should, therefore, be appraised using the principle of proportionality. Employing this principle involves striking a balance between the end the application is attempting to achieve and the means by which it will be realised. This requires a detailed assessment of the intended application and its potential impacts, *i.e.* can it be considered as adequate, relevant and not excessive in those particular circumstances? The analysis should include an examination of the various alternatives, including the non-biometric options, and the financial and technological resources required, in addition to the ethical and legal ramifications of the application in question. When such assessments are based on accurate evidence and valid reasoning and are conducted transparently, the application is more likely to be justified.

In the Council's opinion, the justification of implementing a biometric application is reliant on the application being considered proportionate. Biometric applications should, therefore, be assessed on a case-by-case basis, which involves a consideration of the relevance and necessity of employing biometric technologies, given the proposed purpose of the system, the environment in which it will be used, and the level of efficiency and degree of reliability required to achieve the proposed purpose.

The emergence of mandatory biometric programmes in a number of spheres of public life – ranging from travel and immigration to employment – has the potential to impact on an individual's rights and civil liberties. While recognising that the use of biometric technologies raises particular ethical concerns regarding the rights and interests ordinarily held by all individuals, the possibility still arises that some of these rights may legitimately be limited or overridden where a given biometric application is deemed to be necessary to uphold some common good. Biometric applications are being implemented increasingly by government agencies by appealing to the common good as represented by policies of national and international security, public safety and law enforcement. Governments argue that allowing an individual to opt out of a national biometric programme could impact on the ability of the state to fulfil its responsibility to protect the rights of other citizens. Therefore, while the mandatory enrolment in specific biometric programmes may result in a limiting of a particular individual's right to privacy and autonomy in controlling the use and availability of his/her personal information and the right to opt out, the envisaged improvement in security and safety for everyone is often considered to justify the negative impact at the level of the individual. Nonetheless, deciding when the common good should prevail over an individual's rights is not always evident. The Council is concerned that the common good argument is increasingly being used to justify incursions into people's privacy. The recourse to utilise the common good argument needs to be convincing and based on credible reasoning.

While an individual's rights and civil liberties are deserving of respect and are subject to legal protection, the Council recognises that these rights may be overridden by the state under certain circumstances for the benefit of the common good. However, the Council expresses concern that the argument of upholding the common good may be employed too readily as the reason for implementing particular programmes and applications. Therefore, given the limitations such programmes can place on an individual's civil liberties, there needs to be a proportionate justification and rationale for invoking the common good argument.

While the deployment and use of biometric technologies have increased significantly in recent times, biometrics is still a relatively new concept for many people to fully grasp. Several international studies have indicated that the use of biometric technologies often evokes fears of privacy and civil liberties infringements among the general public. Public acceptance of biometric applications is dependent on the degree of trust in the technology itself and in those operating the applications. In order to encourage increased trust and acceptance, the onus is on system operators to demonstrate both transparency and accountability in the development, implementation and use of biometrics. Facilitating an open and honest discussion between all the relevant stakeholders prior to the implementation of a particular application is an integral part of this trust model. Providing information about the purpose, necessity and proportionality of a given application can help to improve awareness and foster understanding, even among those who may disagree with the application.

The Council believes that increased transparency and honesty regarding biometric technologies, applications, the use to which an individual's biometric information will be put and who will have access to this information is essential in garnering the trust and acceptance of the intended users of these systems. This includes providing information on the most up to date independent research and developments in biometrics and accurate information on the role the biometric application will play in resolving the particular problem at hand. An important aspect of this transparency is the need for a full and frank debate on the issues raised by all parties who will be involved in the proposed application, prior to the establishment of the proposed programme. This is considered particularly important for applications where participation will be mandatory.

Undoubtedly, biometric technologies can provide an accurate and rapid method of identification, thereby enhancing privacy and security – for example, by helping to secure personal information; by assisting an individual to retain control over his/her own information; or by reducing the likelihood of identity theft. However, concerns have been raised regarding the potential for this technology to diminish the level of control an individual has over his/her personal information. These privacy concerns are manifest in two spheres, namely personal privacy (*i.e.* fears about the erosion of personal identity and bodily integrity) and informational privacy, such as fears about the misuse of information and "function creep" (where information collected for a particular purpose is subsequently used for something else).

While defining what exactly privacy is may prove difficult, many people recognise that the right to privacy is of intrinsic importance to them. Many people retain a "sense of privacy", *i.e.* an understanding that certain aspects of their life are no one else's business, but their own. This view is perpetuated through the frequent descriptions of the concept of privacy as an individual's right to be left alone or a barrier against intrusion from the outside world. Privacy facilitates our understanding of our sense of self, *i.e.* the recognition that our bodies, our thoughts and our actions are our own, which is important for the attribution of moral responsibilities. Our ability to control who has access to us and information pertaining to us is closely linked to our ability to form and maintain different types of social relationship with different people. Since our interactions and experiences with other people contribute to our sense of self and of belonging, the concept of privacy is, thus, interconnected with personal identity.

If biometrics were to become the default method of identification there are concerns that this could result in the redefinition of the body as identifying information. The "informatisation" of the body could potentially enable the categorisation and, as a consequence, the discrimination of an individual. Categorisation is often conducted as a method of social control, with people being assigned to different categories, *e.g.* immigrant, suspect, criminal. Ascribing an identity to an individual or labelling him/her in such a way not only has a direct impact on him/her, but also on how he/she is perceived within society. By categorising someone based on their body, it could become difficult for that person to rid him/herself of their assigned identity, even where it is inaccurate or the result of a misidentification.

In addition, stigmatisation and discrimination could also be manifest towards those individuals who wish to use a particular system, but consistently experience problems, *e.g.* due to an injury, a medical condition or some form of disability. Such individuals could, therefore, be excluded from a particular service because they cannot use the "normal" biometric system, resulting in discrimination because their body does not conform to some preset biometric criterion. It is the view of the Council that effective fallback procedures and alternative systems need to be in place to ensure such individuals are not disenfranchised or discriminated against.

The Council recognises the need to establish and/or corroborate the identification of an individual in a globalised world and the many advantages of so doing. However, the method(s) of identification used should in no way be taken to define or categorise a person's identity in a more substantive sense. Indeed, the inappropriate use of bodily information to categorise, stigmatise or discriminate in any way should be resisted strongly. With that in mind, the Council recommends that respect for human dignity should be at the forefront of considerations by policy makers and the biometrics community when designing, implementing and operating biometric technologies and applications.

The informational privacy concerns raised in relation to biometric technologies all stem from the level of control a given individual has over his/her biometric (and other personal) information. The Council see the concept of privacy as a means of controlling a person's personal information, as being interconnected with the notion of bodily integrity and the inviolability of a person's body. Moreover, the inability to control information pertaining to us has implications for our autonomy, dignity and the respect afforded to us as persons.

An individual's biometric information is an intrinsic element of that person. The Council, therefore, recommends that the right to bodily integrity and respect for privacy should apply not only to an individual's body, but also to any information derived from the body, including his/her biometric information.

By providing an inherent link to a given individual's identity, biometric information could potentially be used for numerous different purposes beyond just recognition. This usability of biometric information also increases the opportunity for function creep. Privacy concerns about function creep tend to arise where the purpose of collecting and using the biometric information is not made clear, since the individuals concerned cannot control what their personal information is to be used for. Improvements in the level of interoperability between different biometric systems (which enable greater information sharing) further accentuates these privacy concerns. The Council recognises that suitable protocols are also needed to control the access to, and the use of, databases containing people's biometric and related information.

The Council is of the opinion that, in order to respect and uphold an individual's privacy and confidentiality, biometric applications should utilise only information required to meet a clear, limited and specified purpose. Therefore, any subsequent attempts to use the information for another purpose or to share it with third parties without the knowledge and consent of the individual should be prohibited.

In addition, the Council recommends that appropriate information and access management procedures should be established for all biometric applications to ensure that:

- system operators and system providers are properly trained with regard to their obligations to respect and protect the information;
- system operators and system providers can access only the information they require to conduct their job.

The ability to amalgamate biometric and other personal information can lead to detailed individual profiles being created. Profiling is conducted for a number of different reasons, including for marketing purposes, with the intention of tackling crime, as well as improving public safety and national security. However, questions have been raised regarding the success and, therefore, the justification of these potentially invasive measures.

Profiling invokes fears of discrimination against certain groups within society, for example, through racial profiling, particularly where the operation and management of such measures is not wholly transparent. Such concerns may be alleviated if the rationale for employing the profiling measures is made apparent. In the Council's opinion, the credibility, justification and acceptance of these measures is dependent on the evidence used to support their implementation.

The Council believes that it is essential that profiling measures do not target particular groups within society unfairly or disproportionately. In addition, where an individual is profiled, this should be done in an appropriate manner based on valid reasoning and evidence, and in accordance with due process to ensure that his/her rights and civil liberties are respected and upheld.

The ability of an individual to maintain control over the use of and access to his/her biometric and personal information relates not only to issues of privacy and bodily integrity but also to autonomy. Informed consent is an integral component of exercising one's autonomy. From the Council's perspective, it is imperative that an individual's decision about participating in a biometric application is based on all the details relevant to that application, including what personal information will be collected, the purpose for which it is being collected, how and where the information will be stored, who will have access to it, and whether or not he/she will be able to access, review and amend the stored information.

However, quite apart from being informed of such details, it is important that the individual understands the purpose and implications of the application as well as the potential consequences of his/her own decision to participate or not. Concerns raised in relation to user understanding are particularly relevant for certain potentially vulnerable groups within society, which may not fully appreciate the implications of their participation. Children especially could be vulnerable to being "softened up" to the habitual provision of biometric and other information without being aware of the privacy implications. In the view of the Council, safeguards and procedures need to be in place to ensure that such individuals are protected, but also not disenfranchised. Accordingly, the decisions such individuals make regarding their participation should be facilitated whenever possible, but a parent or legal guardian should be able to act in the person's interest if he/she is not deemed to be competent to make the decision for him/herself.

In order to make the decision whether or not to participate in a biometric programme an individual should be fully and accurately informed and should understand all the issues and implications relating to the provision of his/her information. The Council considers that the issue of user understanding is of particular importance for biometric applications that will be used by potentially vulnerable groups (*e.g.* the elderly, the very young or those with mental and/or learning disabilities). Where such individuals are deemed competent and aware of the consequences of their decision, this decision should be respected. However, if the person is not considered competent, decisions regarding his/her participation should be made by his/her parent or legal guardian. In the case of biometric applications involving children (*i.e.* individuals under 18 years of age), the assent of the child should be sought as well as the consent of his/her parent or legal guardian.

The ability to collect some forms of biometric information covertly means that an individual's consent may not always be sought prior to acquiring his/her information. While such covert collection (*e.g.* surveillance) is usually conducted for the purposes of crime prevention and detection, the Council is of the opinion that there is a limited number of scenarios where such covert collection could be justified. Acquiring someone's information without his/her knowledge, consent or cooperation impinges on his/her privacy and autonomy and often evokes fears of a "Big Brother" type society where everyone is under suspicion. The potential participants of such surveillance activities need to be made aware not only that their biometric information could be collected, but also the reasons for its collection.

Where biometric information is to be collected without an individual's cooperation, the Council considers that, subject to legal exceptions, system operators have an obligation to notify the potential participants (whether willing or unwilling) that the collection of biometric information is ongoing in that area. Moreover, system operations should also provide some explanation as to why the biometric information is being collected and who will have access to it.

In exercising one's autonomy, the principle of informed consent implies that a person's decision whether or not to participate in a given biometric application is voluntary. An individual should, therefore, be entitled to opt out of a biometric programme should he/she so wish. Moreover, when someone opts out of a biometric system, he/she should not be placed at a disadvantage to those who are willing to utilise that system. In the Council's view, the failure to provide non-biometric alternative systems would discriminate against those individuals who are unwilling to provide their biometric information. In addition, non-biometric systems should not be downgraded or neglected as a means of encouraging or coercing people to use biometric systems.

Notwithstanding certain compulsory biometric applications, the Council recommends that an individual should be entitled to exercise his/her autonomy freely and without any external influences when deciding whether or not to enrol in a given application. The Council considers it important that non-biometric alternative systems should be made available, where practicable, for those individuals who do not want to use the biometric system, and individuals should not be disenfranchised or discriminated against by choosing not to participate in a given biometric programme.

When an individual opts to enrol in a given biometric application, he/she provides his/her biometric information for the purposes of being recognised, *i.e.* for verification or identification. However, given the nature of biometric information, it may also be possible to derive additional medical and sensitive personal information from certain biometric identifiers. The possible privacy implications this could have for the individual involved, should such information be used for another purpose, is of concern to the Council.

In line with the *Data Protection Acts* (1988 and 2003), the Council recommends that biometric systems should only collect that information required to fulfil a prescribed purpose. Since the overarching purpose of biometric systems is to verify or identify a given individual, any additional medical or sensitive personal information collected incidentally, which is not needed for recognition purposes, should be deleted from the system.

In the Council's view, the rights of privacy, autonomy and bodily integrity ensure that an individual retains a level of control and ownership over his/her personal information even after this information has been collected and stored. Individuals should, therefore, be able to determine the nature of the information about them being stored. Moreover, an individual should also be entitled to certify that any stored information pertaining to him/her is accurate and up to date. To facilitate such clarifications, system operators need to implement review and audit mechanisms. Such auditing measures would also help to identify information that is no longer relevant or appropriate to continue storing, for example, information relating to someone who has left the biometric programme or has died.

While recommending that an individual should be entitled to access and review information pertaining to him/her, the Council concedes that, under certain circumstances (*i.e.* in the interest of the common good), an individual may be prohibited from accessing this information. Such a situation might arise where the individual's information is necessary to a criminal investigation.

An individual should have the right to access any collected and/or stored information relating to him/her and to review and amend it where necessary, subject to legal exceptions. Moreover, if an individual no longer wishes to utilise the biometric application or the original purpose of the application has been achieved, then any biometric and other personal information about that person should be deleted from the system.

Notwithstanding the recommendations made pertaining to the collection, storage of, and access to biometric information, several technical and practical measures can be implemented to ensure the security and privacy of an individual's biometric information. When enrolling in a biometric system, the salient discriminatory features from an individual's biometric modality (*e.g.* his/her fingerprint) are extracted and used to generate a template, which is a digital, numeric representation of that modality. Using templates, particularly where they are encrypted, as opposed to raw images makes it much more difficult to regenerate the original biometric information, thus, offering greater privacy protection. Biometric systems operate in two basic modes, namely (i) verification and (ii) identification. Verification involves a one-to-one comparison to authenticate an individual's claimed identity, whereas identification involves comparing an individual's template with all the templates in a given database (*i.e.* a one-to-many comparison). Verification-based systems enable an individual to retain control over his/her biometric information because his/her template can be stored locally (*e.g.* on a smart card)[1] and not in a centralised database, unlike identification-based systems. In the Council's opinion, verification-based systems, therefore, provide better safeguards for privacy. While preferring biometric systems that do not utilise centralised databases, the Council acknowledges that such databases may be required for certain applications. However, in

1 A smart card is a card shaped portable data carrying device, which contains a microchip that can be used to both store and process data.

order to counteract threats of data mining[2] and function creep, in such cases, an individual's biometric information should be stored separately from his/her other personal information. Further efforts that can be taken to increase both privacy and security include the use of cryptosystems and biometric encryption, which enables template comparison and matching to be conducted in an encrypted domain (*e.g.* by using a password or key generated from a biometric feature).

The Council recommends that certain technical and practical measures should be established in order to ensure the integrity of an individual's personal and informational privacy. Therefore, subject to justifiable exceptions, templates should be used instead of raw images; applications should be verification-based as opposed to identification-based; systems using databases should store the biometric information separately to other personal information, with these databases being connected by a secure network; and cryptographic systems and biometric encryption should be implemented.

Privacy is a fundamental right that is recognised in many international instruments and regulations, for example, all European countries have enacted legislation safeguarding privacy and Directive 95/46/EC of the European Union (EU) focuses directly on protecting personal data. However, while privacy legislation is well established in most jurisdictions, there is, currently, very little legislation in Europe or further afield which deals specifically with biometric technologies. Concerns have thus been raised in a number of quarters about the ability of existing legislation to provide sufficient protection to biometric information. Consequently, there have been calls, which the Council echoes, for privacy legislation to be reviewed and updated to take account of developments in the use of biometrics and related technologies.

The Council recommends that biometric data should be classified as sensitive personal information and as such afforded greater protection. Consequently, the Council is of the opinion that Ireland's existing data protection legislation does not deal sufficiently with the privacy concerns presented by the increasingly mainstream use of biometrics. The Council welcomes the decision by the Minister for Justice, Equality and Law Reform in November 2008 to establish a committee to review current legislation and urges the committee to consider the privacy/data protection implications arising from biometric technologies.

2 Data mining is data research and analysis aiming to extract hidden trends or correlations from large data sets or to identify strategic information.

CHAPTER 1
AN OVERVIEW OF BIOMETRICS AND ITS APPLICATIONS

Chapter 1: An Overview of Biometrics and its Applications

What are Biometrics?

A biometric is any measurable, physical or physiological feature or behavioural trait that can be used to identify an individual or to verify the claimed identity of an individual.[3,4] Examples of physiological biometrics include fingerprints, hand geometry, the face, the iris, the retina, the venous networks of the hand and even body odour. Behavioural biometrics include voice,[5] signature, keystroke dynamics (manner of typing on a keyboard) and gait (manner of walking).[6]

While the range of body features that can be used for biometric recognition has greatly expanded since this technology was first established, not all physiological or behavioural characteristics are suitable for biometric recognition. In order to be considered suitable for use in biometric recognition, a physiological or behavioural characteristic is usually evaluated against a number of criteria: (i) universality, (ii) distinctiveness, (iii) permanence, (iv) collectability, (v) performance, (vi) acceptability and (vii) resistance to circumvention (see Table 1).[7,8,9,10] These are sometimes referred to as the "seven pillars of biometrics". While no biometric modality fulfils all seven of the pillars equally well, certain modalities satisfy more of the criteria than others (*e.g.* fingerprint and iris would score better overall than dynamic signature and keystroke dynamics) and would, therefore, be deemed more reliable or "stronger" in terms of their suitability for recognition purposes. In addition, for large-scale applications (*e.g.* in airports) high-speed matching is required and this can favour the selection of one particular biometric modality over another. A more detailed review of different biometric modalities is given in Chapter 2.

---

3 European Commission Joint Research Centre, Institute for Prospective Technology Studies (2005). *Biometrics at the Frontiers: Assessing the Impact on Society*. Seville, 166p.

4 Cavoukian A and Stoianov A (2007). *Biometric Encryption: A Positive-Sum Technology that Achieves Strong Authentication, Security AND Privacy*. Information and Privacy Commissioner/Ontario, Toronto, 48p.
Available online at: http://www.ipc.on.ca/images/Resources/bio-encryp.pdf, accessed 6 February 2008.

5 Voice is often considered as both a physiological and a behavioural biometric.

6 These lists of biometric features are for illustrative purposes only and are not considered exhaustive.

7 European Commission Joint Research Centre, Institute for Prospective Technology Studies (2005) *op. cit.*

8 Jain AK, Ross A and Prabhakar S (2004). An Introduction to Biometric Recognition. *IEEE Transactions on Circuits and Systems for Video Technology* 14(1): 4–20.

9 Organisation for Economic Co-operation and Development (OECD), Working Party on Information Security and Privacy (2004). Biometric-Based Technologies. OECD, Paris, 66p.

10 Wayman JL (2000). Fundamentals of Biometric Authentication Technologies. In JL Wayman (ed.) *National Biometric Test Center Collected Works 1997–2000*. Version 1.2. San Jose State University, p.1–20.

Table 1: Seven Pillars for a Biometric Characteristic (Modality)[11,12,13]

Universality	All individuals should have the characteristic.
Distinctiveness	The characteristic should be sufficiently different to distinguish between any two individuals.
Permanence	The characteristic should remain largely unchanged throughout the individual's life.
Collectability	It should be relatively easy for the characteristic to be presented and measured quantitatively.
Performance	Refers to the level of accuracy and speed of recognition of the system given the operational and environmental factors involved.
Acceptability	Refers to an individual's willingness to accept the use of that characteristic for the purpose of biometric recognition.
Resistance to Circumvention	Refers to the degree of difficulty required to defeat or bypass the system.

Why are Biometrics Used?

Traditionally, the identification of an individual or the verification of an individual's claimed identity involved the use of a password, personal identification number (PIN) or cryptographic key ("something you know") or the possession of an identity (ID) card, smart card or token ("something you have").[14,15] However, there are a number of problems associated with these security measures. For example, passwords and PINs can be forgotten, shared with others, and lost or stolen, which could compromise the integrity of the system. A biometric trait is part of an individual and as such it offers the third element of proof of identity, *i.e.* "something you are". Consequently, biometric traits are thought to have a number of advantages over the aforementioned security measures: they cannot be lost or forgotten, they are difficult to copy, forge or share and they require the individual to be present at the time of identification.[16,17,18,19]

The use of biometrics also makes it difficult for an individual to repudiate having accessed a physical location or a computer system, or having conducted a particular transaction. In fact, biometric traits are often portrayed as the ultimate form of identification or verification,[20]

11 European Commission Joint Research Centre, Institute for Prospective Technology Studies (2005) *op. cit.*

12 Jain *et al.* (2004) *op. cit.*

13 Jain AK, Ross A and Pankanti S (2006). Biometrics a Tool for Information Security. *IEEE Transactions on Information Forensics and Security* 1(2): 125–143.

14 European Commission Joint Research Centre, Institute for Prospective Technology Studies (2005) *op. cit.*

15 Jain A, Bolle R and Pankanti S (1999). Introduction to Biometrics. In A Jain, R Bolle and S Pankanti (eds.) *Biometrics: Personal Identification in Networked Society*, Kluwer Press, Dordrecht, p.1–42.

16 Jain *et al.* (2006) *op. cit.*

17 European Commission Joint Research Centre, Institute for Prospective Technology Studies (2005) *op. cit.*

18 Jain *et al.* (1999) *op. cit.*

19 National Science and Technology Council (NSTC) Subcommittee on Biometrics (2006b). *The National Biometrics Challenge.* Washington, 19p. Available online at: http://www.biometrics.gov/Documents/biochallengedoc.pdf, accessed 10 October 2007.

20 Cavoukian and Stoianov (2007) *op. cit.*

and are being promoted in many quarters as a means of heightened security, efficiency and convenience and have been proposed as the solution to issues of identity theft and benefit fraud.[21,22] It is envisaged that biometric systems will be faster and more convenient to use, cheaper to implement and manage and more secure than traditional identification and verification methods.[23,24] Nonetheless, biometrics also have their limitations, for example, passwords, PINs and ID cards can all be re-issued relatively easily if they become compromised, which is not the case for an individual's fingerprint or iris image. The practical and technological aspects and limitations of biometric recognition systems are discussed in more detail below.

Architecture and Design of Biometric Recognition Systems

Architecture and Design of Biometric Recognition Systems

Although humans have been using certain features (*e.g.* face, voice and gait) to recognise each other for thousands of years,[25,26,27] the automated and semi-automated approach used in biometric recognition systems is a relatively recent development from the last few decades.[28] While the mechanisms involved and the modalities (characteristics) used may vary, there are four basic stages in biometric systems: (i) enrolment, (ii) storage, (iii) acquisition and (iv) matching. With any biometric system, the individuals required to use the system need to be enrolled. Biometric data, for example a fingerprint, is collected using a sensor to produce a digital representation of the data. The system then extracts salient discriminatory features (*i.e.* feature extraction) from the digital representation and these features are used to generate a template (*i.e.* a feature data set), which is then linked to the user's identity and stored in the system.[29,30,31] In basic terms the template takes the form of numeric data.[32,33] The next time the individual presents his/her fingerprint to the sensor the sample template that is acquired is compared to the enrolled (stored) template using a mathematical algorithm. If they match the individual is accepted.[34]

---

21 *The Economist* (2003). Biometrics: Too flaky to trust. *The Economist* 4 December 2003. Available online at: http://www.economist.com/opinion/displaystory.cfm?story_id=E1_NNGGNJD, accessed 17 October 2007.

22 European Commission Joint Research Centre, Institute for Prospective Technology Studies (2005) *op. cit.*

23 Wayman (2000) *op. cit.*

24 European Commission Joint Research Centre, Institute for Prospective Technology Studies (2005) *op. cit.*

25 Mordini E and Ottolini C (2007). Body identification, biometrics and medicine: ethical and social considerations. *Annali dell Istituto Superiore di Sanità* 43(1): 51–60.

26 National Science and Technology Council (NSTC) Subcommittee on Biometrics (2006a). *Biometrics History*. Washington, 27p. Available online at: http://www.biometrics.gov/Documents/BioHistory.pdf, accessed 10 October 2007.

27 Jain *et al.* (2004) *op. cit.*

28 NSTC Subcommittee on Biometrics (2006a) *op. cit.*

29 Not all biometric systems are template based, *e.g.* voice recognition biometrics involves the use of models. In addition, some systems may store the original "raw" image of the biometric modality, *e.g.* the fingerprint or facial image.

30 Jain *et al.* (2004) *op. cit.*

31 OECD, Working Party on Information Security and Privacy (2004) *op. cit.*

32 Woodward JD Jr, Webb KW, Newton EM, Bradley M, Rubenson D, Larson K, Lilly J, Smythe K, Houghton B, Pincus HA, Schachter JM and Steinberg P (2001). *Army Biometric Applications. Identifying and Addressing Sociocultural Concerns.* RAND, California, 185p.

33 European Commission Joint Research Centre, Institute for Prospective Technology Studies (2005) *op. cit.*

34 Wayman (2000) *op. cit.*

However, it should be noted that the templates being compared do not have to be exactly the same for the system to provide a match.[35,36] This is a feature of all biometric systems because no two samples of the same biometric from the same person are ever absolutely identical.[37] This phenomenon is known as intra-class (intra-user) variation and it can be caused by differences in a number of factors between both sample collection times, for instance, differences in the ambient conditions, imperfect imaging conditions, changes in the user's biometric characteristic or in the user's interaction with the sensor.[38,39] Matching is, thus, a statistical process, with the algorithm providing a score of the degree of similarity between the two templates being compared, *i.e.* the higher the matching score the more certain the system is that the two templates belong to the same person. The final decision is regulated by a threshold, which determines the margin of error allowed by the algorithm; therefore, the matching score needs to be above the designated threshold.[40] This threshold level can be adjusted by the system operator to meet the needs of a specific application, *i.e.* decreasing it makes the system more tolerant to user variations, whereas increasing it makes the system more secure.

Modes of Biometric Systems

No matter what application a biometric system is used for (*e.g.* commercial, financial, healthcare, security, or law enforcement), these systems have two basic functions, namely *verification* and *identification*. *Verification* is where the biometric system authenticates an individual's claimed identity by comparing the newly collected sample biometric data with the corresponding enrolled template. For *verification*, the enrolled template may be stored locally, for example, on a smart card or token, or on a database. This is what is known as a **one-to-one** comparison. Identity *verification* is a form of positive recognition, the aim of which is to prevent multiple individuals from using the same identity.[41,42]

Identification is where the system attempts to ascertain who an individual is without that individual claiming a particular identity. In this case the sample biometric is compared with all the templates in a given database, *i.e.* a **one-to-many** comparison. *Identification* in this instance is a form of negative recognition, whereby the aim is to prevent an individual from using multiple identities.[43,44] Biometric *identification* is also used to screen people against specific watch lists and databases.[45]

---

35 Jain *et al.* (2004) *op. cit.*
36 Wayman (2000) *op. cit.*
37 Cavoukian and Stoianov (2007) *op. cit.*
38 Jain *et al.* (2006) *op. cit.*
39 Jain *et al.* (2004) *op. cit.*
40 OECD, Working Party on Information Security and Privacy (2004) *op. cit.*
41 Jain *et al.* (2004) *op. cit.*
42 Wayman (2000) *op. cit.*
43 Jain *et al.* (2004) *op. cit.*
44 Wayman (2000) *op. cit.*
45 National Science and Technology Council (NSTC) Subcommittee on Biometrics (2006c). *Biometrics Frequently Asked Questions.* Washington, 25p. Available online at: http://www.biometrics.gov/Documents/FAQ.pdf, accessed 10 October 2007.

Data Storage

Depending on the system being used, the enrolled templates may be either stored in a centralised database or stored locally (decentralised) on to a portable medium, such as a "smart card". The method of storage of the enrolled templates can be influenced by the biometric recognition mode being used, *i.e.* identification- and screening-based applications require a centralised database, whereas verification-based applications can utilise either central or local storage.[46,47,48]

One of the major criticisms of centralised databases is the potential for function creep, *i.e.* where information originally collected for one purpose is subsequently used for another purpose, which raises an array of privacy concerns.[49] These concerns are often raised regarding the surveillance and data mining capabilities of databases. To assuage these concerns, it has been suggested that an individual's personal information (*e.g.* name, address, *etc.*) should be stored separately from his/her biometric information.[50,51,52] A secure network could then be used to link these separate databases. In addition, methods to encrypt (encode) both the personal and the biometric information should be employed to help maintain database security and integrity.[53,54] Notwithstanding privacy issues, decisions on the design of a database can be performance related, *i.e.* very large databases take longer to search, which affects system processing times and overall user throughput. However, databases can be partitioned into smaller subsections, which would accommodate parallel searches, thus not affecting the overall speed of processing, though this may influence system accuracy.[55,56]

The storage of biometric information on portable media, such as smart cards, ensures that the user retains control of his/her own biometric information, therefore, it cannot be used without the user's consent.[57,58,59] With such systems, the template is usually retrieved from the smart card and compared with the individual's live biometric sample.[60,61] In addition to just storing the biometric template, smart cards can also facilitate match-on-card technology, whereby the template plus all the system modules (*e.g.* feature extraction and matching) are all stored and

46 Article 29 Data Protection Working Party (2003). Working Document on Biometrics. European Commission, Brussels, 11p. Available online at: http://ec.europa.eu/justice_home/fsj/privacy/docs/wpdocs/2003/wp80_en.pdf, accessed 1 November 2007.

47 Cavoukian and Stoianov (2007) *op. cit.*

48 Most CM (2004b). Towards Privacy Enhancing Applications of Biometrics. *Digital ID World* June/July 2004: 18–20.

49 Cavoukian and Stoianov (2007) *op. cit.*

50 International Biometric Group (2007a) *BioPrivacy Best Practices.* International Biometric Group, New York. Available online at: http://www.biometricgroup.com/reports/public/reports/privacy_best_practices.html, accessed 15 February 2008.

51 Cavoukian A (1999). *Privacy and Biometrics.* Information and Privacy Commissioner/Ontario, Toronto, 15p. Available online at: http://www.ipc.on.ca/images/Resources/pri-biom.pdf, accessed 6 February 2008.

52 Most (2004b) *op. cit.*

53 Cavoukian and Stoianov (2007) *op. cit.*

54 International Biometric Group (2007a) *op. cit.*

55 Wayman (2000) *op. cit.*

56 European Commission Joint Research Centre, Institute for Prospective Technology Studies (2005) *op. cit.*

57 *ibid.*

58 Woodward *et al.* (2001) *op. cit.*

59 Cavoukian and Stoianov (2007) *op. cit.*

60 European Commission Joint Research Centre, Institute for Prospective Technology Studies (2005) *op. cit.*

61 Cavoukian and Stoianov (2007) *op. cit.*

conducted on the card and the biometric information never leaves the card.[62,63] However, smart cards can be lost or stolen; therefore, the information they contain should also be encrypted to guard against unauthorised access. In some cases, information stored on the smart card may also be stored in a database as a form of backup and/or to check for counterfeit cards.[64]

Biometric information, whether stored in a database or portable medium, can take the form of encrypted (or unencrypted) templates or "raw" images. Encrypted templates offer greater security as it can be extremely difficult, though not necessarily impossible, to reconstruct the original biometric image from them.[65,66,67,68] In addition, the use of templates instead of raw images can also help to alleviate concerns surrounding the derivation of additional sensitive information (*e.g.* health and medical information) from the collected biometric information. The storage of raw images facilitates interoperability between different biometric systems utilising the same modality. For example, the international standards for machine readable travel documents proposed by the International Civil Aviation Organisation (ICAO) call for the storage of full images to facilitate comparison and identification of individuals with other databases.[69] The storage of raw biometric images also prevents an application operator from being "locked in" to a particular system or vendor product owing to the proprietary nature of the feature extraction, template generation and matching algorithms used. Therefore, raw images can be inputted into another system without the need to re-enrol the users. However, the storage of raw images requires stringent security measures to limit possible abuse of the biometric information.

Finally, biometric systems, whether incorporating centralised or localised information storage, require appropriate information management mechanisms and access controls to be implemented to reduce the likelihood of problems arising, for instance, identity theft, fraud, *etc.*[70,71,72] It has been suggested that such mechanisms should also include policies for the retention and discarding of biometric images and/or templates and the associated personal information.[73,74]

---

62 Jain AK, Nandakumar K and Nagar A (2008). Biometric Template Security. *EURASIP Journal on Advances in Signal Processing. Special Issue Advanced Signal Processing and Pattern Recognition Methods for Biometrics* Volume 2008, Article ID 579416, 17p.

63 Snijder M (2007). *Report on the Workshop Security & Privacy in Large Scale Biometric Systems.* European Biometrics Forum, Dublin, 28p.

64 Wayman (2000) *op. cit.*

65 European Commission Joint Research Centre, Institute for Prospective Technology Studies (2005) *op. cit.*

66 Jain AK, Ross A and Uludag U (2005). Biometric Template Security: Challenges and Solutions. *Proceedings of the European Signal Processing Conference (EUSIPCO '05), Antalya, Turkey, September 2005,* 4p.
Available online at: http://biometrics.cse.msu.edu/Publications/SecureBiometrics/JainRossUludag_TemplateSecurity_EUSIPCO05.pdf, accessed 31 March 2008.

67 Jain *et al.* (2008) *op. cit.*

68 NSTC Subcommittee on Biometrics (2006c) *op. cit.*

69 International Civil Aviation Organization Technical Advisory Group (ICAO TAG) (2004). *Biometrics Deployment of Machine Readable Travel Documents.* ICAO TAG MRTD/NTWG, Technical Report, Version 2, 60p.

70 European Commission Joint Research Centre, Institute for Prospective Technology Studies (2005) *op. cit.*

71 Article 29 Data Protection Working Party (2003) *op. cit.*

72 Data Protection Commissioner (2008a). *Annual Report of the Data Protection Commissioner 2007.* Brunswick Press Ltd, Dublin, 88p.
Available online at: http://www.dataprotection.ie/documents/annualreports/AR2007En.pdf, accessed 13 May 2008.

73 International Biometric Group (2007a) *op. cit.*

74 Data Protection Commissioner (2008a) *op. cit.*

Practical Considerations

System Accuracy and Error Rates

Biometric recognition is a statistical process, which is affected by intra-class variations between enrolment and subsequent acquisitions. Therefore, unlike password- or PIN-based systems, which are either correct or not, no biometric recognition system is 100 per cent accurate and all biometric systems are susceptible to a number of different errors,[75,76,77] for example, failure to enrol, failure to acquire, false accept error and false reject error.[78] The *failure to enrol rate* (FTE) reflects the difficulty (or inability) an individual might have in enrolling in the system to begin with. This can be due to a quality control feature of the system, where poor quality images are rejected.[79,80] Since the enrolment process directly influences the accuracy, efficiency and usability of a biometric system this error is an important consideration. The *failure to acquire rate* (FTA) refers to the difficulty (or inability) of collecting an individual's biometric information during subsequent uses of the system. While the probability of FTE and FTA are quite low, it is considered important to have a fallback procedure or some level of flexibility to cope with these eventualities.[81] For example, an individual could try enrolling another finger, or if the system is multimodal, perhaps he/she could enrol a different modality altogether, for example, his/her face instead of a fingerprint. Human intervention can also aid system flexibility in such instances.

A *false accept error* occurs when an acquired template from one individual, who is not in the system, is mistakenly matched to an enrolled template from another individual.[82,83] False accept errors can compromise the security and integrity of the system. A *false reject error* occurs when an acquired template from one individual does not match the enrolled template for that individual.[84,85] False reject errors are inconvenient to legitimate users of the system, who have to re-attempt the recognition process again or have to be authorised through an alternative mechanism, for example, through human intervention.

There are error rates associated with each of these errors – the *false accept rate* (FAR) and the *false reject rate* (FRR).[86] The FAR and the FRR are inversely proportional, *i.e.* decreasing FAR results in an increased FRR and *vice versa*. The FAR and FRR are influenced directly by the decision threshold of the system, which itself is a function of a particular application.[87]

---

75 Cavoukian and Stoianov (2007) *op. cit.*

76 Jain *et al.* (2004) *op. cit.*

77 Jain *et al.* (2006) *op. cit.*

78 Biometric systems utilising partitioned databases are also prone to binning error, which is where an individual's enrolled template is placed in a different "bin" (partition) to his/her subsequent samples.

79 Wayman (2000) *op. cit.*

80 Jain *et al.* (2004) *op. cit.*

81 NSTC Subcommittee on Biometrics (2006c) *op. cit.*

82 Cavoukian and Stoianov (2007) *op. cit.*

83 European Commission Joint Research Centre, Institute for Prospective Technology Studies (2005) *op. cit.*

84 Cavoukian and Stoianov (2007) *op. cit.*

85 European Commission Joint Research Centre, Institute for Prospective Technology Studies (2005) *op. cit.*

86 The false accept rate (FAR) is also known as the false match rate (FMR) and the false reject rate (FRR) is also known as the false non-match rate (FNMR).

87 The FTE is also linked to FAR and FRR because if the FTE is high, the system will only contain high quality templates, which can help to decrease FAR and FRR overall.

Therefore, there needs to be some trade-off between the two error rates for each application. For example, for high security applications a low FAR is required – therefore, the decision threshold would be set quite high and, as a consequence of the stringent system requirements, the FRR increases. Most biometric systems have FRR ranges from 0.1 per cent to 20 per cent (depending on the modality used and the application), which means that a legitimate user of the system will, on average, be rejected between one in every 1000 times and one in every five times they use the system.[88] FAR rates tend to range from one in 100 for low security applications to approximately one in 10 million for very high security applications.[89] The FAR and the FRR for a given biometric system can be plotted against each other on a curve[90] to indicate system performance at all operating points (thresholds). The point where FAR and FRR are equal is known as the equal error rate (EER). The EER is generally considered to be the best operating level for civilian biometric applications.[91,92]

These FAR and FRR error rates are tested to evaluate overall performance and accuracy and the results are used to promote particular algorithms and biometric systems. While these results are useful initially, the majority are conducted in laboratory environments and therefore may not be an accurate reflection of system performance in "real world" situations.[93,94] Therefore, post-deployment testing and fine tuning are critical if the system is to obtain the performance levels observed in the laboratory when operated in the real world. In addition, test results are often specific for a particular application and will not necessarily translate to other applications using the same modality. Nonetheless, large-scale independent tests have and are being conducted on a number of biometric modalities and the Information and Privacy Commissioner of Ontario (IPC) has recommended that these results should be consulted prior to system implementation.[95,96]

Biometric System Security Vulnerabilities and Countermeasures

Vulnerabilities

The increased acceptance and diffusion of biometric systems will be partially dependent on the perceived security and accuracy of these systems.[97] It is well documented that biometric systems and technologies are vulnerable to both intrinsic failures and failures due to external attacks. Such system failures can threaten security, erode an individual's privacy or deny legitimate users of a particular service. Intrinsic failures are associated with the overall system recognition performance, *i.e.* system errors (FTE, FTA, FAR and FRR), which an adversary could

---

88 Cavoukian and Stoianov (2007) *op. cit.*

89 *ibid.*

90 The curve of FAR plotted against FRR is the receiver operating characteristic (ROC) curve.

91 European Commission Joint Research Centre, Institute for Prospective Technology Studies (2005) *op. cit.*

92 Jain *et al.* (1999) *op. cit.*

93 Wayman (2000) *op. cit.*

94 Jain *et al.* (1999) *op. cit.*

95 Cavoukian and Stoianov (2007) *op. cit.*

96 Independent tests include the Face Recognition Vendor Test (FRVT), the Fingerprint Verification Competition (FVC) and the Iris Challenge Evaluation (ICE).

97 Jain *et al.* (2008) *op. cit.*

utilise to his/her advantage.[98,99] Adversary attacks are intentional efforts to access or circumvent the system illegitimately through the use of vulnerabilities in the design system.[100] Adversary attacks include the following:[101,102,103,104]

Spoofing – where the adversary uses a fake biometric (*e.g.* fake fingerprint, facial image) to fool the system.

Replay attacks – where the adversary records an image from a legitimate user and inserts it back into the system.

Substitution attacks – where the adversary accesses a stored template and overwrites it or replaces it with his/her own template.

Tampering – where the adversary modifies the feature sets in stored templates or during verification to ensure a high match score is achieved for his/her own biometric.

Masquerade attacks – where the adversary creates a digital artefact from the template that is sufficient to produce a match. The created artefact does not, necessarily, have to resemble the original image to ensure a match.

Trojan horse attacks – where the adversary replaces parts of the system, such as the matching algorithm, with a Trojan horse computer program that always produces a high matching score.

Overriding the yes/no response – the output of biometric systems is a binary yes/no response, therefore, the adversary could insert a false yes response to bypass the biometric system.

Countermeasures

Various mechanisms have been developed to help counter-attack and overcome these vulnerabilities. For example, liveness detection[105] and/or human supervision can limit the opportunities for spoofing.[106,107,108] Where biometric information is being transferred across a network, encryption methods and physical security measures can be used to decrease the prospect of attack. Moreover, using secure computer code and software protocols can reduce the likelihood of Trojan horse attacks.[109] As mentioned above, the use of match-on-card

---

98 An individual can utilise these intrinsic failures to gain illegitimate access to the system, without making any effort to circumvent the system, *i.e.* a zero-effort attack.

99 Jain *et al.* (2008) *op. cit.*

100 *ibid.*

101 Cavoukian and Stoianov (2007) *op. cit.*

102 Jain *et al.* (2008) *op. cit.*

103 Acharya L (2006). *Biometrics and Government.* The Parliamentary Information and Research Service, Library of Parliament, Canada, 19p.

104 Jain *et al.* (2005) *op. cit.*

105 Liveness detection is a method of checking if the biometric sample is being read from a live person as opposed to a fake body part or the body part from a dead person. Liveness detection can involve checking different physiological signs such as blood pressure, pulse rate, respiration, skin conductivity or temperature.

106 International Biometrics Group (2007c). *Liveness Detection in Biometric Systems.* International Biometric Group, New York. Available online at: http://www.biometricgroup.com/reports/public/reports/liveness.html, accessed 15 February 2008.

107 Jain *et al.* (2008) *op. cit.*

108 Snijder (2007) *op. cit.*

109 Jain *et al.* (2008) *op. cit.*

technology can help to increase security, but this technology is expensive and is not considered appropriate for large-scale applications.[110] Furthermore, an adversary could attempt to access the information on a card if it was lost or stolen.

The possibility of biometric templates being compromised is seen as a major problem, because it has been shown that biometric templates can be reverse engineered to produce the original image or an approximation of it.[111,112] Since an individual's biometric characteristic itself cannot be revoked or reissued, it is crucial that mechanisms to protect stored templates are devised and implemented.[113] For this reason, there has been a lot of research into cancellable (revocable) biometric templates, *i.e.* if a template is compromised it can be revoked and a new template can be generated without the need to re-collect the biometric sample.[114] This technology uses a mathematical transformation function, defined by a randomly generated password or PIN, to transform the biometric template.[115,116] The new "transformed" template is then stored in the system.[117,118] The same transformation function is applied to each sample template before comparison with the stored template. If a transformed template is compromised, it is cancelled and a new transformation function is generated and applied to the original biometric to produce a newly transformed template.[119,120,121] This technology also enables an individual to use the same biometric modality, for instance, his/her iris, in several different applications because the transformation function will be different in each case.

Biometric cryptosystems are another countermeasure against biometric system insecurities and vulnerabilities. This technology combines biometrics and cryptography to facilitate biometric matching in the encrypted domain, which enhances system security.[122,123] In a conventional cryptosystem the user inputs a simple PIN or password to unlock an encryption key, for instance, a series of numbers, which is then used to decode the information the user wishes to access. In a simplified biometric cryptosystem, the user's biometric template is used instead of a password to unlock the encryption key.[124] A more secure cryptosystem would be to generate the encryption key directly from the biometric template; however, such a system is difficult to design.[125,126] Due to intra-class variation no two biometric templates are exactly the same,

---

110 *ibid.*

111 Jain *et al.* (2005) *op. cit.*

112 International Biometrics Group (2007b). *Generating Images from Templates.* International Biometric Group, New York. Available online at: http://www.biometricgroup.com/reports/public/reports/templates_images.html, accessed 15 February 2008.

113 Jain *et al.* (2008) *op. cit.*

114 Jain *et al.* (2005) *op. cit.*

115 Teoh ABJ and Yuang CT (2007). Cancellable Biometrics Realization with Multispace Random Projections. *IEEE Transactions on Systems, Man, and Cybernetics – Part B: Cybernetics* 37(5): 1096–1106.

116 Jain *et al.* (2008) *op. cit.*

117 *ibid.*

118 The transformation may be classed as invertible or non-invertible, though, with non-invertible forms it is difficult, but not impossible to recover the original template.

119 Teoh and Yuang (2007) *op. cit.*

120 Jain *et al.* (2008) *op. cit.*

121 Cavoukian and Stoianov (2007) *op. cit.*

122 Jain *et al.* (2005) *op. cit.*

123 Cavoukian and Stoianov (2007) *op. cit.*

124 *ibid.*

125 Jain *et al.* (2008) *op. cit.*

126 European Commission Joint Research Centre, Institute for Prospective Technology Studies (2005) *op. cit.*

which would mean that the encryption key would be slightly different every time the user tried to access the system; therefore, access would be denied.[127,128] However, the company GenKey has managed to overcome this issue and has developed an algorithm that can generate a cryptographic key from data extracted from fingerprint presentations.[129,130]

An alternative system is "biometric encryption": when an individual enrols his/her biometric sample (*e.g.* his/her fingerprint), the system randomly generates a digital key (which is unknown to the user and the system operator). The encryption algorithm then binds this key securely to the biometric sample to create an encryption template.[131] The key and the biometric sample are then discarded and they cannot be regenerated from the stored encryption template. When the individual next uses the system, he/she provides a biometric sample and if it matches the biometric template the encryption algorithm retrieves the key. This key can then be used in a conventional cryptosystem.[132] At the end of the recognition process the biometric is again discarded; thus, the system stores only the biometrically encrypted key, not the biometric itself, which should help to alleviate privacy concerns.[133,134] Since the key is independent of the biometric, it can be revoked if the system is compromised and a new key could be generated, which offers increased flexibility and security. Unfortunately, such systems are difficult to design and there is currently only one commercially available biometric encryption system, *i.e.* the Philips priv-ID™ system.[135,136]

Standards and Interoperability

Different vendors tend to use proprietary algorithms (*e.g.* for feature extraction and matching) in their biometric systems. As a result, a biometric template generated in one system, using one specific algorithm, would not normally be suitable for use with another system, *i.e.* the systems are not interoperable.[137,138,139] Consequently, an individual user would have to enrol separately in each different system even if the systems were based on the same biometric modality.

Standards for biometric systems and modalities have and are being developed to try to overcome interoperability issues and to promote open systems.[140,141,142] For example, the

---

127 Cavoukian and Stoianov (2007) *op. cit.*

128 Jain *et al.* (2008) *op. cit.*

129 Thieme M (2008) *Business Potential for Genkey Technology.* International Biometric Group, New York. Available online at: http://genkeycorp.com/index.php?n=19&task=vis&id=5, accessed 15 February 2008.

130 More information is available online at: http://genkeycorp.com/what-we-do/Performance/, accessed 15 February 2008.

131 Cavoukian and Stoianov (2007) *op. cit.*

132 *ibid.*

133 *ibid.*

134 European Commission Joint Research Centre, Institute for Prospective Technology Studies (2005) *op. cit.*

135 Cavoukian and Stoianov (2007) *op. cit.*

136 Cherry S (2007). Personal biometrics, private data. *Password, Philips Research Technology Magazine* 30: 5–8.

137 OECD, Working Party on Information Security and Privacy (2004) *op. cit.*

138 Huijgens R (2006) Technology trends – 2006. In PE Schmitz, R Tavano, J Lodge, R Huijgens, K Aisola and M Flammang (eds.). *Biometrics in Europe. Trend Report 2006.* Unisys Corporation, Brussels, 113p.

139 ICAO TAG (2004) *op. cit.*

140 OECD, Working Party on Information Security and Privacy (2004) *op. cit.*

141 National Science and Technology Council (NSTC), Subcommittee on Biometrics (2006d). *Privacy & Biometrics: Building a Conceptual Foundation.* Washington, 57p. Available online at: http://www.biometrics.gov/docs/privacy.pdf, accessed 10 October 2007.

142 Snijder (2007) *op. cit.*

Minutiae Template Interoperability Testing (MTIT) Project is a major European initiative that was undertaken to test and improve the interoperability of minutiae-based fingerprint systems.[143] The US National Institute for Standards and Technology (NIST) has defined the Common Biometric Exchange File Format (CBEFF), which provides a storage format to facilitate the exchange of data between systems and organisations.[144,145,146] In addition, the International Civil Aviation Organization (ICAO) Technical Advisory Group has devised standards for machine readable travel documents, for example, visas or e-passports, which state, among other stipulations, that all biometric travel documents must contain an image of the document holder's face;[147] this image must be of a particular storage size and format; the image must be stored on a remotely readable (*i.e.* contactless chip); and the actual document should be valid for ten years.[148]

Part of the challenge of interoperability is to enable the easy, rapid and seamless integration of system components into already functioning systems and then interchange these components as necessary without compromising the functionality of the system.[149] It is also important that standards are developed in a particular format to facilitate the interoperability of new systems with legacy (*i.e.* already existing) information or new information with legacy systems.[150] According to the US National Science and Technology Council (NSTC) Subcommittee on Biometrics, the benefits of interoperability include "real-time, controlled and documented data sharing between biometric systems; consistent enterprise-wide performance across different user groups and organizations; integration of disparate systems produced by different vendors; eradication of non-operability caused by proprietary middleware, hardware and software".[151] Interoperability between different systems is important both at a national level, for example, between different government-based applications,[152] and internationally as is the case with machine readable travel documents. Therefore, testing and evaluation of biometric systems is required to ensure both vendor and intersystem compliance with and conformity to the requisite standards.[153]

---

143 For more information see the Minutiae Template Interoperability Testing Project website: http://www.mtitproject.com/, accessed 8 July 2009.

144 OECD, Working Party on Information Security and Privacy (2004) *op. cit.*

145 Dessimoz D, Richiardi J, Champod C and Drygajlo A (2006). *Multimodal Biometrics for Identity Documents (MBioID): State-of-the-Art* (Version 2.0), Research Report, PFS 341–08.05, Institut de Police Scientifique – Ecole des Sciences Criminelles (Université de Lausanne) & Speech Processing and Biometric Group (Ecole Polytechnique Fédérale de Lausanne), 156p.
Available online at: http://www.europeanbiometrics.info/images/resources/90_264_file.pdf, accessed 7 February 2008.

146 For further examples of existing standards in biometrics see Dessimoz *et al.* (2006) *op. cit.*

147 These facial images may be supplemented with fingerprint and/or iris images.

148 ICAO TAG (2004) *op. cit.*

149 NSTC Subcommittee on Biometrics (2006b) *op. cit.*

150 National Science and Technology Council (NSTC) Subcommittee on Biometrics and Identity Management (2007). NSTC Policy for Enabling the Development, Adoption and Use of Biometric Standards. Washington, 11p.
Available online at: http://www.biometrics.gov/Standards/NSTC_Policy_Bio_Standards.pdf, accessed 2 May 2008.

151 NSTC Subcommittee on Biometrics (2006b) *op. cit.*

152 *ibid.*

153 NSTC Subcommittee on Biometrics and Identity Management (2007) *op. cit.*

The Council recommends that certain technical and practical measures should be established in order to ensure the integrity of an individual's personal and informational privacy. Therefore, subject to justifiable exceptions, templates should be used instead of raw images; applications should be verification-based as opposed to identification-based; systems utilising databases should store the biometric information separately to other personal information, with these databases being connected by a secure network; and cryptographic systems and biometric encryption should be implemented.

Technology Associated with Biometric Systems

Radio Frequency Identification

While radio frequency identification (RFID) technology is not a form of biometrics, it can be integrated into biometric systems.[154] This technology is based on the use of RFID tags (transponders) and readers (transceivers), which communicate with each other using radio frequency signals.[155,156,157] RFID tags are either passive or active: passive tags do not have their own power supply and derive their energy from the radio waves transmitted by the reader; active tags contain their own battery and can generate their own radio waves. RFID tags can be used in two basic ways. Firstly, the RFID tag might store only information that can be accessed using the reader (*i.e.* a "read-only" system).[158,159] Secondly, an alternative "read-write" system allows information to be added to or deleted from the RFID tag – however, this requires greater processing power for the tag.[160,161] One main advantage of this technology is that the information stored on an RFID tag can be read remotely in a contactless system, without a line of sight to the reader.[162,163] The maximum operating distance (*i.e.* the range) between the RFID reader and the tag varies from a few centimetres to tens of metres. This range depends on a number of factors such as the frequency being used, the power of the reader, sources of radio interference and objects in the environment that might reflect or absorb radio waves.[164,165] While a contactless system offers more flexibility, measures need to be taken to prevent unauthorised access to (*i.e.* skimming), and/or tampering with, the information passed to and from the RFID tag, thus maintaining the security and integrity of the system.[166,167]

154 NSTC Subcommittee on Biometrics (2006c) *op. cit.*

155 *ibid.*

156 Hodges S and McFarlane D (2005). *Radio frequency identification: technology, applications and impact.* Auto-ID Labs White Paper Series, Edition 1. Available online at: http://www.autoidlabs.org/single-view/dir/article/6/60/page.html, accessed 8 April 2008.

157 Parliamentary Office of Science and Technology (POST) (2004). Radio Frequency Identification (RFID). *Postnote* 225: 1–4.

158 Hodges and McFarlane (2005) *op. cit.*

159 POST (2004) *op. cit.*

160 Hodges and McFarlane (2005) *op. cit.*

161 POST (2004) *op. cit.*

162 Hodges and McFarlane (2005) *op. cit.*

163 POST (2004) *op. cit.*

164 Hodges and McFarlane (2005) *op. cit.*

165 POST (2004) *op. cit.*

166 Nygren S (2007) Non-contact and RFID – not only in logistics. *Detektor International* 3: 14–15.

167 Hodges and McFarlane (2005) *op. cit.*

The small size of RFID tags enables them to be inserted into a multitude of different objects, for instance, identity cards, credit cards, swipe cards and library books,[168] and also into animals, for example, pets and livestock. Once inserted, this technology can be used to identify the object/animal, track its location, or, in the case of a library book, store information about who has borrowed the book, or, in the case of an animal, to store information relating to the animal's owner. Moreover, these tags can also be implanted into people to track their location (*e.g.* individuals on parole from prison, Alzheimer's patients), to monitor particular medical conditions remotely, or for the individual's convenience (*e.g.* for electronic payments or as a type of club membership card[169]).

In terms of biometric systems, RFID tags are predominantly used to store an individual's biometric information, for example, in a smart card or e-passport, from where it can be retrieved to verify that individual's identity by comparison with their live biometric sample (see Figure 1).[170]

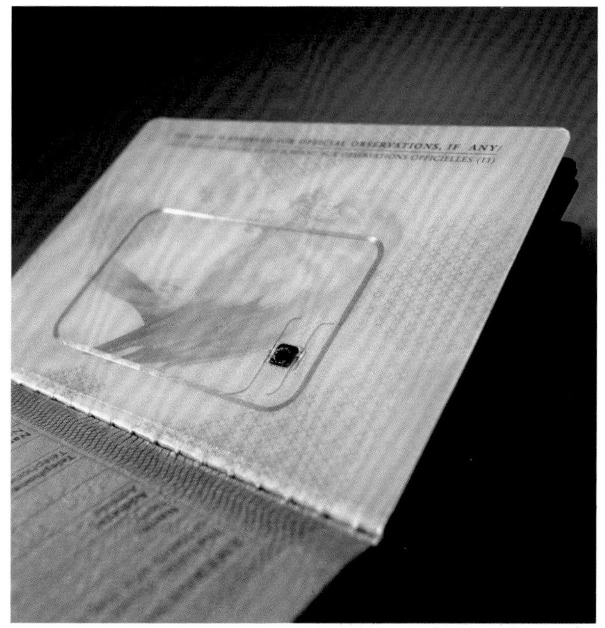

Steve Horrell/Science Photo Library

Figure 1: A biometric passport showing an RFID chip. The chip is a tiny RFID computer chip, which stores information (including biometric information) about the holder in electronic form.

168 RFID tags can also be inserted into membership cards, consumer goods, and electrical products, *etc.*

169 BBC News (2004). *Barcelona clubbers get chipped.* BBC News, published online 29 September 2004. Available online at: http://news.bbc.co.uk/2/hi/technology/3697940.stm, accessed 8 April 2008.

170 Nygren S (2007) *op. cit.*

CHAPTER 2
OVERVIEW AND COMPARISON OF BIOMETRIC MODALITIES

Chapter 2: Overview and Comparison of Biometric Modalities

Market Overview

Biometric systems, whether used for verification or identification, can be employed in numerous different contexts, for example, security, surveillance and law enforcement, e-commerce, e-government and physical and logical access. The use of biometric systems is expected to increase even further in the future,[171,172] with revenues in the biometrics industry expected to grow from approximately $3.4 billion in 2009 to over $9 billion by 2015.[173] Currently, biometric systems based on fingerprint recognition comprise the majority of the biometrics market with Automated Fingerprint Identification Systems (AFISs) accounting for 33–38 per cent while non-automatic fingerprint recognition systems make up between 25 and 28 per cent of the market.[174] Face recognition systems account for between 11 and 18 per cent of the market,[175,176,177] with most of the remainder of the market split between iris recognition (5.1–7.0 per cent), hand geometry (1.8–7.0 per cent), voice recognition (3–5 per cent), vein recognition (2.4–3.0 per cent), multimodal biometrics (2.9 per cent) and a mixture of other modalities (1.6–4.0 per cent).[178,179] Given the potential security, efficiency and convenience that such systems are purported to provide, the diffusion of biometric technologies to all aspects of society is projected to increase; however, this increase will not be uniform across all biometric modalities. For example, iris-, voice- and multimodal-based recognition systems are expected to show the greatest level of increase, whereas it is anticipated that traditionally strong market sectors such as hand geometry and fingerprint recognition will decrease somewhat.[180]

The current dominance of fingerprint modalities in the biometrics market does not mean that fingerprints are considered the optimum biometric, but is more a result of the maturity of this technology.[181,182] In fact, it is widely considered that there is no optimal biometric that can satisfy the requirements of all types of application. All biometrics have particular advantages and disadvantages that need to be assessed, and the final decision on which biometric system

171 European Commission Joint Research Centre, Institute for Prospective Technology Studies (2005) *op. cit.*

172 Biever C (2005). ID revolution – prepare to meet the new you. *New Scientist* 187(2516): 26–29.

173 International Biometric Group (2008). Available online at: http://www.biometricgroup.com/reports/public/market_report.php, accessed 8 April 2009.

174 Most CM (2007). *Mega Trends and Meta Drivers for the Biometrics Industry: 2007–2020.* Presentation at the Biometrics Exhibition and Conference 2007, 17–19 October 2007, Westminster, London.

175 International Biometric Group (2008) *ibid.*

176 Biometric Technology Today (2007b). Face Recognition: Part 2. *Biometric Technology Today* 15(10): 10–11.

177 Most CM (2007) *op. cit.*

178 International Biometric Group (2008) *op. cit.*

179 Most CM (2007) *op. cit.*

180 *ibid.*

181 Biever (2005) *op. cit.*

182 O'Gorman L (1999). Fingerprint Verification. In A Jain, R Bolle and S Pankanti (eds.) *Biometrics: Personal Identification in Networked Society*, Kluwer Press, Dordrecht, p.43–64.

to implement depends on the environment and the application.[183,184,185] However, this decision can be informed by comparing and evaluating each biometric on the basis of the criteria outlined in Table 1 above (universality, distinctiveness, permanence, collectability, performance, acceptability and resistance to circumvention) and the relevance of these criteria to the proposed application. In addition to these criteria, another factor that should be considered when implementing a particular biometric system is the cost–benefit analysis of one particular biometric compared to another.[186]

Fingerprint Recognition

Current Situation

Fingerprint-based systems are the most commonly used of any biometric recognition system. These technologies occupy over 50 per cent of the biometrics market and this dominance is linked to their adaptability and flexibility for use in numerous different applications. As noted previously, fingerprint recognition technologies comprise both AFIS and non-AFIS technologies, but there are substantial differences between these two types of recognition system.[187,188,189] For the purposes of this document, unless otherwise stated, fingerprint recognition will refer to non-AFIS.

Basic Information

The skin on the surface of a fingertip consists of raised folds of skin, known as ridges, and these ridges are separated by valleys. The pattern of ridges and valleys on a fingertip represents a fingerprint, which is what is used in biometric recognition.[190,191,192] Biometric fingerprint recognition involves the comparison of specific major features and/or minor fingerprint features. The three major fingerprint features used for pattern recognition are arches, loops and whorls, one of which is found on a given fingerprint (see Figure 2).[193,194] Two other major features may also be used for recognition, *i.e.* the core and delta. The core is the centre point of a particular fingerprint pattern and the delta is a point from which three patterns

183 Jain *et al.* (2004) *op. cit.*

184 NSTC Subcommittee on Biometrics (2006c) *op. cit.*

185 Most CM (2003). Battle of the Biometrics. *Digital ID World* November/December 2003: 16–18.

186 European Commission Joint Research Centre, Institute for Prospective Technology Studies (2005) *op. cit.*

187 AFIS technology is used for law-enforcement purposes and involves the comparison of fingerprints obtained from a possible suspect against a database of criminal/crime scene fingerprints. With this system, all ten fingerprints from a given individual are collected and compared with the database. Any potential matches are verified by expert fingerprint examiners – therefore, this form of recognition is semi-automated and not carried out in real time. Non-AFIS recognition is fully automated and is conducted in real time, and does not necessarily require all ten fingerprints to be used.

188 National Science and Technology Council (NSTC) Subcommittee on Biometrics (2006e). *Biometrics Overview.* Washington, 10p. Available online at: http://www.biometricscatalog.org/NSTCSubcommittee/Documents/Biometrics%20Overview.pdf, accessed 10 October 2007.

189 European Commission Joint Research Centre, Institute for Prospective Technology Studies (2005) *op. cit.*

190 O'Gorman (1999) *op. cit.*

191 NSTC Subcommittee on Biometrics (2006e) *op. cit.*

192 Jain *et al.* (2004) *op. cit.*

193 Loops make up nearly two thirds of all fingerprints, whorls comprise almost one third, with approximately 5–10% of fingerprints consisting of arches, see International Biometric Group (2007d). *Fingerprint Feature Extraction.* International Biometric Group, New York. Available online at: http://www.biometricgroup.com/reports/public/reports/finger-scan_extraction.html, accessed 18 April 2008.

194 OECD, Working Party on Information Security and Privacy (2004) *op. cit.*

deviate.[195,196] The core and the delta can be used as landmarks to orient two fingerprints for matching; however, it should be noted that these features are not found on all fingerprints.[197,198]

The minor features used in fingerprint recognition are known as minutiae – hence, the process is known as minutiae matching. Minutiae are discontinuities that disrupt the flow of fingerprint ridges and there are two main types, *i.e.* endings and bifurcations. An ending is where a ridge stops and a bifurcation is where a ridge splits in two (see Figure 3).[199,200,201,202]

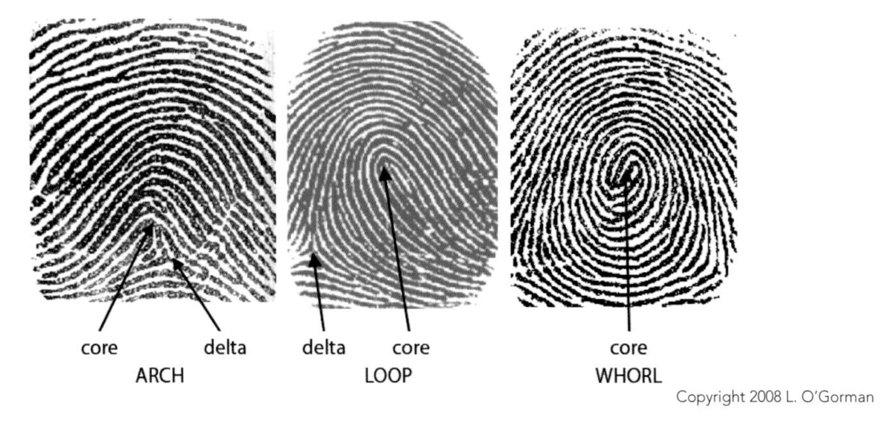

Figure 2. Fingerprint patterns: arch, loop and whorl. Fingerprint landmarks, core and delta are also shown. Note there is no delta shown in the whorl image.[203]

Figure 3. Fingerprint minutiae – ending and bifurcation.[204]

195 O'Gorman (1999) *op. cit.*

196 International Biometric Group (2007d) *op. cit.*

197 O'Gorman (1999) *op. cit.*

198 Although features such as scars and creases could potentially be used for recognition purposes, they are not normally used because they may be transient or even introduced artificially.

199 O'Gorman (1999) *op. cit.*

200 International Biometric Group (2007d) *op. cit.*

201 European Commission Joint Research Centre, Institute for Prospective Technology Studies (2005) *op. cit.*

202 Fingerprint ridges also consist of pores at regular intervals, which have been researched for recognition purposes, but this methodology requires very high resolution images.

203 This image is reproduced with the permission of Dr Lawrence O'Gorman, Bell Laboratories Research, Murray Hill, New Jersey, US.

204 This image is reproduced with the permission of Dr Lawrence O'Gorman, Bell Laboratories Research, Murray Hill, New Jersey, US.

The basic steps of fingerprint recognition are generally the same for both pattern and minutiae matching. A high quality image is initially collected using one of three different sensor types, namely optical, silicon (capacitance) or ultrasound. With an optical sensor the user places their finger on the sensor surface (platen) and a laser light illuminates the fingerprint. This light is reflected by the ridges of the fingerprint and is converted to a digital signal.[205,206]

The process of feature extraction is a crucial step in fingerprint recognition since all subsequent operations are dependent on the quality of the image.[207,208] While the actual feature extraction algorithm used is proprietary to the system vendor, the general process involves reducing the "noise" of the image and enhancing the ridge definition to allow more precise detection of the minutiae (or pattern features).[209,210,211] The algorithm filters out distortions and false minutiae caused by dirt, scars, sweat, *etc.*,[212] but this may also result in the deletion of actual minutiae.[213,214] The resulting template contains between 10 and 100 minutiae,[215] whereas the original image would have contained between 50 and 200 minutiae.[216] Approximately 80 per cent of vendors utilise minutiae matching in some format, with the remainder using pattern matching. In pattern matching, the enrolled template represents a series of ridges from the original fingerprint, and during verification (or identification) this is compared with a submitted template corresponding to the same area of the fingerprint. The use of multiple ridges reduces the dependency on individual minutiae points, which can be affected by wear and tear, during the matching process.[217,218,219] However, as a result, pattern matching templates tend to be larger than minutiae templates, 900–1200 bytes compared with 250–700.[220,221,222]

---

205 The oils on the fingerprint ridges form a seal with the platen and this reflects the light.

206 O'Gorman (1999) *op. cit.*

207 European Commission Joint Research Centre, Institute for Prospective Technology Studies (2005) *op. cit.*

208 International Biometric Group (2007d) *op. cit.*

209 O'Gorman (1999) *op. cit.*

210 International Biometric Group (2007d) *op. cit.*

211 European Commission Joint Research Centre, Institute for Prospective Technology Studies (2005) *op. cit.*

212 O'Gorman (1999) *op. cit.*

213 To facilitate comparison between an enrolled template and a sample template, each minutia point is assigned a number of feature attributes including the type of minutia, its location, its direction, and often its distance from the core.

214 International Biometric Group (2007d) *op. cit.*

215 O'Gorman (1999) *op. cit.*

216 OECD, Working Party on Information Security and Privacy (2004) *op. cit.*

217 International Biometric Group (2007d) *op. cit.*

218 OECD, Working Party on Information Security and Privacy (2004) *op. cit.*

219 Dessimoz *et al.* (2006) *op. cit.*

220 International Biometric Group (2007d) *op. cit.*

221 OECD, Working Party on Information Security and Privacy (2004) *op. cit.*

222 Dessimoz *et al.* (2006) *op. cit.*

Applications of Fingerprint Recognition

Physical and logical access[223] control as well as employee time and attendance monitoring are the most widespread applications for fingerprint recognition systems. However, the small sensor size, allied to the adaptability of fingerprint recognition systems has resulted in this technology being built into numerous other devices and products including laptop computers, USB (universal serial bus) storage devices, cars, household door locks, safes, and even mobile phones.[224,225,226]

In the financial and retail sectors these systems have been implemented to enable users to pay for goods and services or to access bank ATMs (automatic teller machines) using their fingers.[227,228]

Fingerprint recognition systems are also being used in schools, for example, in the US (United States of America) and the UK (United Kingdom), to serve numerous functions, such as securing access to the school grounds, monitoring student attendance and controlling access to student cafeterias and library accounts.[229,230,231] Irish schools have piloted this technology for controlling access and attendance, with the aim of improving administrative efficiency and reducing student truancy.[232,233]

Notwithstanding the automatic fingerprint identification type systems (AFIS) used in law enforcement, other large-scale governmental applications have been implemented including national identification cards, voter registration schemes, management of social service benefits (*e.g.* to counteract fraud) and particularly border control and immigration programmes. For example, when travelling to an increasing number of countries, including the US[234] or Japan, the vast majority of foreign nationals have their fingerprints collected (as well as their photograph taken). In addition, the electronic border control system (e-Channel) operating in Hong Kong uses fingerprint recognition technology. Each Hong Kong resident has a smart ID card, which contains a template of their thumb and other personal information. Users insert their smart cards into the reader at the first gate. The stored data is retrieved from the card and they pass to the second gate, where they

---

223 Logical access refers to the process of accessing or logging on to a computer, network or database.

224 European Commission Joint Research Centre, Institute for Prospective Technology Studies (2005) *op. cit.*

225 OECD, Working Party on Information Security and Privacy (2004) *op. cit.*

226 The Economist (2006). Biometrics gets down to business. *The Economist* 30 November 2007. Available online at: http://www.economist.com/science/tq/displaystory.cfm?story_id=E1_RPTNNQG, accessed 17 October 2007.

227 *ibid.*

228 Most CM (2004a). Biometrics and Financial Services – Show Me the Money! *Digital ID World* January/February 2004: 20–23.

229 OECD, Working Party on Information Security and Privacy (2004) *op. cit.*

230 European Commission Joint Research Centre, Institute for Prospective Technology Studies (2005) *op. cit.*

231 Becta (2007). *Becta guidance on biometric technologies in schools.* Becta, UK, 10p. Available online at: http://schools.becta.org.uk/upload-dir/downloads/becta_guidance_on_biometric_technologies_in_schools.doc, accessed 5 November 2007.

232 Data Protection Commissioner (2008a) *op. cit.*

233 Yongo I (2007). Technology used to combat truancy. *The Irish Times* 25 April 2007.

234 Department of Homeland Security (2004). *US-VISIT Program Privacy Policy.* Washington, 5p. Available online at: http://www.dhs.gov/xlibrary/assets/privacy/privacy_stmt_usvisit.pdf, accessed 11 April 2008.

scan their thumb. If the thumb scan matches the template on the smart card, the user passes through the gate.[235]

The Eurodac system, which has been operational since 2003, was implemented as a means of comparing the fingerprints of asylum seekers and illegal immigrants throughout the EU (European Union) to determine which Member State is responsible for examining an asylum application.[236,237] Each individual Member State establishes its own fingerprint recognition system, which is then connected to the central EU database. The fingerprints of any asylum applicant entering an EU Member State are screened against the database to establish whether or not that individual has already claimed asylum in another Member State or if he/she has previously entered the EU illegally. If a match is found the individual concerned can be deported from the country he/she is currently attempting to enter back to the country where he/she originally tried to seek asylum. The Eurodac system was implemented to help overcome problems of "asylum shopping", whereby an individual whose application for asylum is rejected in one EU country moves to a different country and applies again.[238] The use of fingerprints helped to overcome the problems of individuals applying in different countries under different names, with fake or unreliable identification documents or no documentation at all. An interesting aspect of the Eurodac system is that the information in the central database is stored anonymously, linked to a specific reference number.[239,240]

Critical Profile of Fingerprint Recognition

The formation of fingerprint patterns is determined during the first seven months of foetal development[241,242] and this process is influenced by the development of nerves in the skin.[243] These patterns are unique for each individual, including identical twins, and also for the fingerprints on each finger of the same person.[244,245] In addition to uniqueness, universality is also an important factor for biometric recognition and almost all individuals have fingerprints apart from those with certain hand-related disabilities and injuries. However, it has been estimated that, at any given time, approximately 4–5 per cent of the population have

235 Wong R (2007). *Innovative use of biometrics in Hong Kong*. Presentation at the Biometrics Exhibition and Conference 2007, 17–19 October 2007, Westminster, London.

236 Eurodac Supervision Coordination Group (2007). *EURODAC Supervision Coordination Group, Report of the first coordinated inspection*. European Data Protection Supervisor, Brussels, 16p.

237 European Commission Joint Research Centre, Institute for Prospective Technology Studies (2005) *op. cit.*

238 Council Regulation (EC) No 2725/2000 of 11 December 2000 concerning the establishment of 'Eurodac' for the comparison of fingerprints for the effective application of the Dublin Convention.

239 *ibid.*

240 Most (2004b) *op. cit.*

241 Jain *et al.* (2004) *op. cit.*

242 European Commission Joint Research Centre, Institute for Prospective Technology Studies (2005) *op. cit.*

243 Matsumoto T, Matsumoto H, Yamada K and Hoshino S (2002). Impact of Artificial "Gummy" Fingers on Fingerprint Systems. *Proceedings of SPIE* 4677: 275–289.

244 Jain *et al.* (2004) *op. cit.*

245 O'Gorman (1999) *op. cit.*

fingerprints that are unusable for recognition purposes.[246,247,248] This problem is, at least partially, a function of the permanence of fingerprints. In the majority of cases fingerprints are stable and long lasting, but they can be damaged due to ageing (thin skin gives poor print resolution), environmental factors (humidity, water, sweat, dirt, chemicals), and occupational factors (manual labour can result in worn down prints, cut and bruised skin).[249,250,251]

Damaged and worn fingerprints affect performance directly because they can result in poor quality images, which provide less information from which to generate a template and thus conduct a comparison. In some cases, the image quality could be so poor that the user may not be able to enrol into the system in the first place or they might need to re-enrol using another finger. Image quality can also be influenced by the type of sensor used – for example, certain sensors can cope better with dirty fingers, thin skin or worn fingerprints.[252,253,254] Notwithstanding these problems, fingerprint recognition offers very good performance and accuracy, particularly for verification systems and small- to medium-scale identification systems (up to a few 100 users). In the case of large-scale, identification-based systems, recognition performance may decrease somewhat because the inherent intra-class variation is magnified resulting in more recognition errors (FAR and FRR) and because the likelihood of FTE and FTA increases with a larger user population. However, in such cases performance can be improved through the enrolment and comparison of multiple fingers from the same person.[255,256,257]

Fingerprint sensors are relatively easy to use,[258,259] requiring minimal training of and feedback to the user, all of which enhances the collectability of fingerprints. The convenience of using the sensors also facilitates the acceptance of fingerprint technologies, though there are still some concerns related to the association of fingerprints with criminality.[260,261,262] There are also some health and hygiene concerns regarding dirt on sensors and the possible transmission of bacteria, but it has been stated that fingerprint sensors are no more unhygienic than a door handle or hand rail.[263,264] In addition, regular sensor cleaning and the advent of non-contact sensors can lessen such hygiene issues.

---

246 Woodward *et al.* (2001) *op. cit.*

247 Jain *et al.* (2004) *op. cit.*

248 European Commission Joint Research Centre, Institute for Prospective Technology Studies (2005) *op. cit.*

249 Jain *et al.* (2004) *op. cit.*

250 Huijgens (2006) *op. cit.*

251 NSTC Subcommittee on Biometrics (2006c) *op. cit.*

252 O'Gorman (1999) *op. cit.*

253 International Biometric Group (2007e). *Optical – Silicon – Ultrasound.* International Biometric Group, New York. Available online at: http://www.biometricgroup.com/reports/public/reports/finger-scan_optsilult.html, accessed 6 March 2008.

254 Rowe RK (2005). A Multispectral Sensor for Fingerprint Spoof Detection. *Sensors* 22(1): 1–4.

255 O'Gorman (1999) *op. cit.*

256 Jain *et al.* (2004) *op. cit.*

257 OECD, Working Party on Information Security and Privacy (2004) *op. cit.*

258 NSTC Subcommittee on Biometrics (2006c) *op. cit.*

259 OECD, Working Party on Information Security and Privacy (2004) *op. cit.*

260 European Commission Joint Research Centre, Institute for Prospective Technology Studies (2005) *op. cit.*

261 OECD, Working Party on Information Security and Privacy (2004) *op. cit.*

262 Woodward *et al.* (2001) *op. cit.*

263 European Commission Joint Research Centre, Institute for Prospective Technology Studies (2005) *op. cit.*

264 NSTC Subcommittee on Biometrics (2006c) *op. cit.*

With regard to circumventing fingerprint recognition systems, a number of studies have shown that these systems can be spoofed using artificial fingers and fingerprints made from readily available materials (*e.g.* gelatine, silicon and Play Doh™), or even cadaver fingers.[265,266,267] In fact, instructions on how to spoof specific sensors are freely available on the Internet. A number of approaches can and have been implemented to reduce the susceptibility of these systems to attack. For example, efforts in liveness detection can overcome some spoof attacks involving artificial or dead fingers, with sensors able to detect various traits including temperature, pressure, electrical measurements, perspiration, pulse oximetry, blood pulsation, skin distortion, odour (of synthetic materials), and organic versus synthetic material or non-living flesh.[268,269,270] The recognition of *multiple* fingers from a given individual makes it more difficult to spoof the system. Recognition of multiple fingers also facilitates the implementation of challenge response systems – for instance, the user could be asked to scan different fingers in a specific sequence; therefore, an attacker will need to have artificial fingers or prints for all of the possible enrolled fingers. Finally, adequate human supervision can be a major deterrent against spoof attacks.[271]

Palm Print Recognition

Basic Information

The palms of the hand have patterns of ridges and valleys, similar to those found in fingerprints, which can be used for biometric recognition.[272,273] These systems use a number of different sensor types, *i.e.* optical, capacitance, ultrasound and thermal.[274] Depending on the resolution of the sensor, the captured images can contain all the features of the palm including the ridge and valley features, the principal lines and wrinkles, as well as hand geometry measurements.[275] Similarly to fingerprint recognition the systems extract minutiae and/or pattern details, which are used to create a template.[276] The template can be representative of the entire palm surface or it can be confined to specific smaller regions of the palm surface, depending on the performance requirements. The matching process can involve minutiae-based matching, correlation-based matching or ridge-based matching.[277] The use of palm print recognition technology is increasing in commercial and law enforcement applications.

---

265 Parthasaradhi STV, Derakhshani R, Hornak LA and Schuckers SAC (2005). Time-Series Detection of Perspiration as a Liveness Test in Fingerprint Devices. *IEEE Transactions on Systems, Man, and Cybernetics – Part C: Applications and Reviews* 35(3): 335–343.

266 Matsumoto *et al.* (2002) *op. cit.*

267 Huijgens (2006) *op. cit.*

268 Parthasaradhi *et al.* (2005) *op. cit.*

269 Antonelli A, Cappelli R, Maio D and Maltoni D (2006). Fake Finger Detection by Skin Distortion Analysis. *IEEE Transactions on Information Forensics and Security* 1(3): 360–373.

270 Mordini E and Massari S (2008). Body, Biometrics and Identity. *Bioethics* 22(9): 488–498.

271 International Biometrics Group (2007c) *op. cit.*

272 Jain *et al.* (2004) *op. cit.*

273 National Science and Technology Council (NSTC) Subcommittee on Biometrics (2006f). *Palm Print Recognition.* Washington, 10p. Available online at: http://www.biometricscatalog.org/NSTCSubcommittee/Documents/Palm%20Print%20Recognition.pdf, accessed 10 October 2007.

274 *ibid.*

275 Jain *et al.* (2004) *op. cit.*

276 NSTC Subcommittee on Biometrics (2006f) *op. cit.*

277 *ibid.*

Critical Profile of Palm Print Recognition

Similarly to fingerprints, palm prints are not universal and they are susceptible to the same problems of wear and tear. However, because a palm represents a larger area than a fingerprint, these features are considered to be even more distinctive than fingerprints.[278] In addition, the minutiae characteristics of palms are more distinctive than the ridge characteristics.[279] The collection of palm prints can be assisted through user feedback, for example, regarding the positioning of the hand. Palm print recognition is considered to be highly accurate, though the quality of the images can affect the error rates. Minutiae-based matching is more accurate than correlation-based matching, but it can take longer.[280] However, system speed can be assisted by partitioning the database into different sections. It has been suggested that palm print recognition accuracy will improve with further technological advances, though independent testing will be needed to corroborate these results.

From a practical perspective, palm print sensors are larger and, consequently, more expensive than fingerprint sensors.[281] Decisions to implement palm print recognition systems must balance the need for accuracy against the cost and the interoperability issues associated with this technology.[282]

Hand Geometry

Hand geometry recognition systems have been commercially available since the 1970s and 1980s.[283,284] The original systems were mostly introduced for physical access control, time and attendance, and the use of hand geometry recognition systems has not altered much from these basic functions.

Basic Information

Hand geometry recognition systems measure the physical dimensions of a hand (or finger) from a 3D image.[285,286] The measurements collected include the shape, width and length of the fingers and knuckles, and the thickness of the hand (or finger).[287,288,289] The user places his/her hand on the sensor, which includes guidance poles to ensure the correct positioning of the user's hand and fingers. The sensor uses a camera to take images of both the top and the side

278 Jain *et al.* (2004) *op. cit.*

279 NSTC Subcommittee on Biometrics (2006f) *op. cit.*

280 *ibid.*

281 Jain *et al.* (2004) *op. cit.*

282 NSTC Subcommittee on Biometrics (2006f) *op. cit.*

283 NSTC Subcommittee on Biometrics (2006a) *op. cit.*

284 Zunkel RL (1999). Hand Geometry Based Verification. In A Jain, R Bolle and S Pankanti (eds.) *Biometrics: Personal Identification in Networked Society*, Kluwer Press, Dordrecht, p.87–101.

285 The same technology is used for finger geometry systems.

286 Zunkel (1999) *op. cit.*

287 Jain *et al.* (2004) *op. cit.*

288 OECD, Working Party on Information Security and Privacy (2004) *op. cit.*

289 Zunkel (1999) *op. cit.*

of the hand (see Figure 4).[290,291] The sensor does not record any surface details, such as finger or palm prints, scars, or skin colour and the resulting image is black and white.[292]

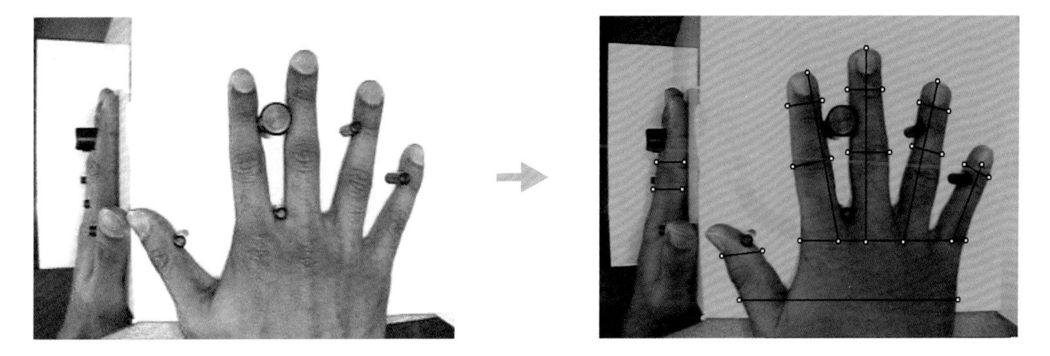

Figure 4: Collecting a hand geometry sample.[293]

During enrolment, the sensor takes almost 100 measurements of the user's hand.[294,295] This may be done up to three times, before the algorithm involved generates a template from the average of these three measurement occasions.[296,297] The resulting template is a mathematical representation of the measurements that were taken. In some cases during enrolment, the enrolee may be required to type in an ID number or PIN (*e.g.* an employee number), which is linked to the template for future verification.[298,299] When the individual next uses the sensor he/she is required to type in his/her ID number or PIN to claim the stored identity. The individual then scans his/her hand and if it is sufficiently similar to the stored template, the system generates a match. The templates generated using hand geometry systems are usually quite small (usually 10 bytes or less),[300] which facilitates their storage on smart cards (or other portable media).[301] This provides two advantages: (i) reducing the amount of computational resources required for the system and (ii) lessening privacy concerns relating to the storage of personal and biometric information in a central database.

290 OECD, Working Party on Information Security and Privacy (2004) *op. cit.*

291 National Science and Technology Council (NSTC) Subcommittee on Biometrics (2006g). *Hand Geometry*. Washington, 7p. Available online at: http://www.biometrics.gov/Documents/HandGeometry.pdf, accessed 10 October 2007.

292 Zunkel (1999) *op. cit.*

293 This image is taken from Ross A, Jain A and Pankanti S. *A Hand Geometry-Based Verification System*. Available online at: http://biometrics.cse.msu.edu/hand_proto.html, accessed 12 August 2008. This image is reproduced with the permission of the authors.

294 NSTC Subcommittee on Biometrics (2006g) *op. cit.*

295 Zunkel (1999) *op. cit.*

296 NSTC Subcommittee on Biometrics (2006g) *op. cit.*

297 Zunkel (1999) *op. cit.*

298 NSTC Subcommittee on Biometrics (2006e) *op. cit.*

299 Zunkel (1999) *op. cit.*

300 OECD, Working Party on Information Security and Privacy (2004) *op. cit.*

301 Zunkel (1999) *op. cit.*

Applications of Hand Geometry Recognition

Hand geometry recognition systems are primarily used for physical access control, time and attendance in workplaces and schools, though they can also be used for point of sale transactions in schools, hotels, athletic and fitness clubs.[302,303]

In terms of a large-scale application, the most well known programme was the US Immigration and Naturalization Service's Passenger Accelerated Service System (INSPASS). This system enabled frequent travellers to quickly pass through immigration at a number of international airports in the US and Canada.[304,305] When a user enrolled in the system he/she received a smart card, containing a template of his/her hand scan, which could be used at designated kiosks.[306] At the kiosks, each user rescanned his/her hand for comparison with the template on the smart card. Over 60,000 individuals were enrolled in INSPASS before its suspension in 2004, due to changes in the border control and immigration policies post-September 11th 2001.[307,308]

Critical Profile of Hand Geometry Recognition

As noted above, hand geometry is quite a popular biometric technology, which is related to the ease of use of these systems, *i.e.* collectability is high.[309,310] The sensor is straightforward to use and requires only limited training to ensure correct hand placement, which is further facilitated by the guidance poles.[311] Such training, particularly if provided during enrolment, reduces problems for subsequent interaction with the sensor, thus increasing the likelihood of obtaining good quality images.[312] In addition to high usability and collectability, hand geometry also offers a good degree of universality since the vast majority of people, apart from those with particular injuries (*e.g.* amputation) and medical conditions (*e.g.* arthritis), possess scannable hands.[313] Hand geometry is generally considered to be stable, particularly once an individual reaches adulthood.[314,315] Moreover, newer hand geometry sensors can "learn" of minor changes in the size of the hand that might be associated with growth or ageing and

302 OECD, Working Party on Information Security and Privacy (2004) *op. cit.*

303 Zunkel (1999) *op. cit.*

304 OECD, Working Party on Information Security and Privacy (2004) *op. cit.*

305 United States Immigration and Naturalization Service (INS), Office of Inspections. *INS Passenger Accelerated Service System (INSPASS) Briefing Paper.* Available online at: http://www.biometrics.org/REPORTS/INSPASS2.html, accessed 11 April 2008.

306 *ibid.*

307 NSTC Subcommittee on Biometrics (2006a) *op. cit.*

308 OECD, Working Party on Information Security and Privacy (2004) *op. cit.*

309 Acharya (2006) *op. cit.*

310 Jain *et al.* (2004) *op. cit.*

311 OECD, Working Party on Information Security and Privacy (2004) *op. cit.*

312 Zunkel (1999) *op. cit.*

313 An individual can be enrolled to use a right-handed sensor with their left hand (with the palm facing upwards) and vice versa, which may help to alleviate some universality and collectability constraints.

314 Zunkel (1999) *op. cit.*

315 NSTC Subcommittee on Biometrics (2006c) *op. cit.*

update the template accordingly.[316] This process of template averaging allows the system to adapt to slow changes in hand geometry without requiring the user to re-enrol.

The acceptability of hand geometry systems is quite high owing to their ease of use, non-invasive nature and the fact that they are mostly used in the verification mode, which limits the amount of information relating to a given user that is stored in the system. Similarly to fingerprint recognition, hygiene issues have been raised due to multiple users, but again these can be alleviated through regular cleaning of the sensor. Cleaning also improves the optical path, which can aid performance.[317] Overall, hand geometry systems offer high performance in the verification mode and in small-scale identification applications, but they are not considered to be distinctive enough for large-scale applications.[318] Also while hand geometry systems generally work well both indoors and outdoors, the performance can be adversely affected by environmental factors such as sunlight and extreme cold.[319] They are also vulnerable to circumvention and spoofing using artificial hands, but this problem can be reduced through adequate supervision and certain methods of liveness detection.

From a practical perspective, hand geometry sensors are relatively bulky, despite the size decreases associated with newer silicon-based systems. In addition, this larger size also prevents these sensors from being embedded in other devices, for example, laptops. However, they can be linked to existing systems, including door locking mechanisms and time and attendance systems.

Vein Pattern Recognition

Basic Information

In vein pattern recognition systems a high resolution camera and infrared light are used to capture the pattern and structure of blood vessels visible on the back of an individual's hand or finger (see Figure 5).[320,321] The algorithm registers the vascular pattern characteristics (*e.g.* blood vessel branching points, vessel thickness and branching angles) and stores these as a template for comparison with subsequent samples from the enrolled individual.[322,323] This technology has the potential to be linked with existing recognition systems such as fingerprint and palm recognition sensors. Vein pattern recognition systems are increasingly being used in order to access ATM cash dispensers and banking services, and for physical access to hospitals

316 Zunkel (1999) *op. cit.*

317 *ibid.*

318 Jain *et al.* (2004) *op. cit.*

319 Zunkel (1999) *op. cit.*

320 National Science and Technology Council (NSTC) Subcommittee on Biometrics (2006h). *Vascular Pattern Recognition.* Washington, 6p. Available online at: http://www.biometrics.gov/Documents/VascularPatternRec.pdf, accessed 10 October 2007.

321 Huijgens (2006) *op. cit.*

322 NSTC Subcommittee on Biometrics (2006h) *op. cit.*

323 Huijgens (2006) *op. cit.*

and universities as well as for residential access, particularly in Japan.[324,325] These recognition systems are also being used for high security network access and in point of sale terminals.

Andrew Brookes, National Physical Laboratory/Science Photo Library

Figure 5: A sensor scanning the vein pattern from the back of an individual's hand.

Critical Profile of Vein Pattern Recognition

The random pattern of blood vessels under the skin is relatively distinct and stable, thus enabling its use for some forms of biometric recognition.[326] Sensors are non-contact and relatively easy to use, though additional guidance brackets may be used to facilitate correct hand positioning.[327] Images cannot be collected at a distance and since the systems are non-contact no latent images are left behind after sensing, which encourages acceptance.[328,329] System performance is quite accurate and because it is difficult to counterfeit blood vasculature, vein recognition is seen as a secure biometric modality.[330]

---

324 NSTC Subcommittee on Biometrics (2006h) *op. cit.*

325 Huijgens (2006) *op. cit.*

326 *ibid.*

327 Huijgens R (2007). Trends in biometrics technology. In PE Schmitz, R Huijgens and M Flammang (eds.). *Biometrics in Europe. Trend Report 2007*. Unisys Corporation, Brussels, 39p.

328 NSTC Subcommittee on Biometrics (2006h) *op. cit.*

329 Huijgens R (2007) *op. cit.*

330 NSTC Subcommittee on Biometrics (2006h) *op. cit.*

Facial Recognition

Current Situation

The face is the biometric characteristic that is most commonly used by people to identify each other.[331,332] Biometric facial recognition has experienced strong growth in recent years and the future growth of this area will be influenced by a number of major policy decisions, particularly those relating to the introduction of biometric identity cards, electronic passports (e-passports) and other machine readable travel and identity documents.[333,334]

Basic Information

Biometric facial recognition is an automated or semi-automated process, which records and compares the spatial geometric distinguishing features of the face.[335] This can include the location and shape of facial attributes – including the eyes, eyebrows, nose, lips, chin – and their spatial relationships, analysis of the entire facial images, and even the analysis of skin texture.[336] Research in this field has been ongoing since the 1960s,[337] and multiple approaches have been devised, which are based on two dimensional (2D) and three dimensional (3D) images and even infrared facial scans.[338,339,340] During enrolment, a sensor (*e.g.* a camera) captures an image or series of images of the user's face, which is converted to a digital format (see Figure 6).[341] An algorithm then extracts the relevant features and measurements and creates a template, which is much smaller than the original image, *i.e.* between 100 and 3500 bytes for the template and 20–40 kilobytes for the original image.[342,343] This small template size facilitates the storage of templates, for example on a smart card or an e-passport, which can then be used for verification-based applications. On each subsequent use of the system, the features of the face are compared between the enrolled template and the sample template and if these are considered sufficiently similar, *i.e.* above a designated threshold, the system records a match. However, in some identification-based applications the matching software may provide a series of potential matches, which are then checked by a human operator before a final decision is made. It should be noted that the ICAO standards for machine readable travel documents insist that the raw image of an individual's face be stored on the chip, within the document, whereas template storage is optional.[344]

---

331 European Commission Joint Research Centre, Institute for Prospective Technology Studies (2005) *op. cit.*

332 Adler A and Schuckers ME (2007). Comparing Human and Automatic Face Recognition Performance. *IEEE Transactions on Systems, Man, and Cybernetics – Part B: Cybernetics* 37(5): 1248–1255.

333 Biometric Technology Today (2007b) *op. cit.*

334 European Commission Joint Research Centre, Institute for Prospective Technology Studies (2005) *op. cit.*

335 Woodward *et al.* (2001) *op. cit.*

336 Jain *et al.* (2006) *op. cit.*

337 NSTC Subcommittee on Biometrics (2006e) *op. cit.*

338 European Commission Joint Research Centre, Institute for Prospective Technology Studies (2005) *op. cit.*

339 Bowyer KW, Chang KI, Flynn PJ and Chen X (2006). Face Recognition Using 2-D, 3-D, and Infrared: Is Multimodal Better than Multisample? *Proceedings of the IEEE* 94(11): 2000–2012.

340 Infrared facial scans are discussed in the section on facial thermography.

341 For some applications a normal photograph can be used for enrolment, *e.g.* with e-passports.

342 Note 1 kilobyte is equal to 1000 bytes.

343 OECD, Working Party on Information Security and Privacy (2004) *op. cit.*

344 ICAO TAG (2004) *op. cit.*

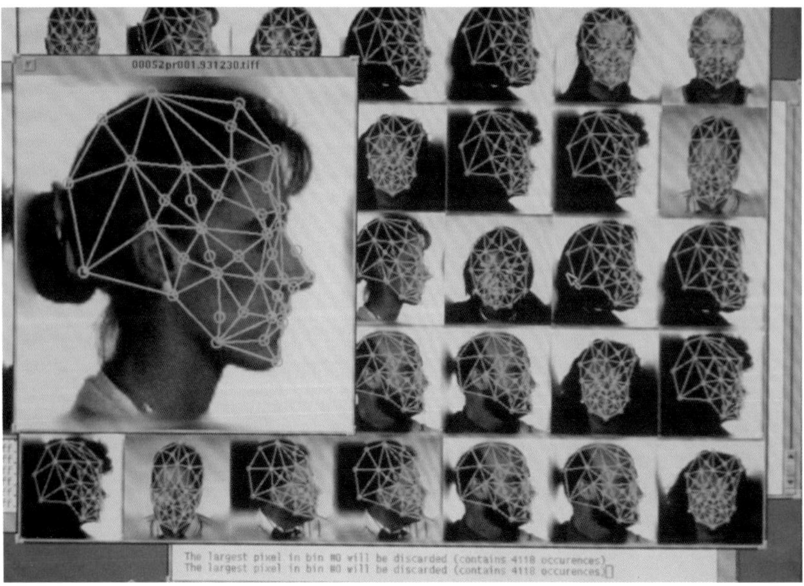

Volker Steger, Peter Arnold Inc./Science Photo Library

Figure 6: A profile of a woman's head with an overlaid contour map used in computerised face recognition. The digital face recognition programme works by comparing the relative distances and directions between specific points on a person's head.

Applications of Facial Recognition

Facial recognition systems are used for physical and logical access control in numerous settings, for example, banks, casinos, offices, crèches, *etc.*, as well as to access computer systems. Facial recognition has also been employed to prevent unauthorised exit from certain locations. For example, a nursing home in Hong Kong has installed facial recognition software to prevent vulnerable patients (*e.g.* with Alzheimer's disease) from leaving the nursing home unaccompanied. If they attempt to leave, they are recognised by the system, which locks the doors and alerts staff.[345]

The traditional use of photographs in identity cards and passports has been expanded to incorporate facial recognition systems. For example, since 2005 all new passports issued in Ireland have been biometric passports, in conjunction with the ICAO standards and also to maintain Ireland's participation in the US Visa Waiver Program. Biometric passports (e-passports) contain a small integrated chip (a radio frequency identification [RFID] chip), embedded in the photo page, which contains a digitised image of the photograph on the passport, as well as all the additional biographical information

345 Kurniawan SH (2007). *Bringing convenience and security to everyday life – Case Histories: A1 Team Malaysia and Elderly Home Association*. Presentation at the Biometrics Exhibition and Conference 2007, 17–19 October 2007, Westminster, London.

visible on the passport.[346] The digitised facial image stored in the passport can then be used in conjunction with facial recognition software to confirm the identity of the passport holder. Due to the use of these passports in facial recognition systems. stringent guidelines have been implemented regarding the quality of the photographs individuals need to submit when applying for their passports. Currently, there are no plans to store fingerprints on Irish passports, but the potential is there and such a situation could arise in future if required internationally.[347]

Finally, the potential to collect facial images at a distance has facilitated the use of this technology for surveillance purposes and for screening individuals against watch lists, with varying degrees of success.[348,349,350]

Critical Profile of Facial Recognition

Faces are universal – therefore, enrolment is always possible.[351] Moreover, facial images are easy to collect: for instance, 2D recognition requires only a photograph and the necessary sensors (cameras) are widely available. It is also possible to capture facial images from a distance, depending on the resolution of the camera involved. The convenient, non-contact, non-invasive mode of collection, allied to the fact that people are accustomed to being recognised by their facial appearance, facilitates the high level of acceptance of facial recognition systems. However, concerns have been raised surrounding the potential to capture facial images covertly, through surveillance equipment, without the knowledge and/or consent of the subject.[352,353] Religious and cultural sensitivities relating to the exposure of the face, particularly for women, may also need to be addressed to encourage the acceptance of facial recognition systems. However, these issues could be addressed through the use of segregated screening areas, staffed by all female attendants.

Unfortunately, facial characteristics exhibit limited variation and this lack of distinctiveness presents challenges for biometric recognition systems, particularly for large-scale, identification-based applications.[354] The impermanence of facial features is also problematic in terms of recognition. For example, features can change dramatically as a result of injury, ageing, cosmetic surgery, or substantial weight gain or weight loss. These changes can be significant enough to require an individual to re-enrol in the system. Many identification documents, for example, passports and drivers licences, often require renewal (with a new

---

346 For more information see the Department of Foreign Affairs (http://dfa.ie/home/index.aspx?id=3030) and the Citizens Information Board (http://www.citizensinformation.ie/categories/travel-and-recreation/travel-abroad/passports-and-visas-to-travel-abroad/machine_readable_irish_passports_and_travel_to_the_US/?searchterm=biometrics), accessed 6 November 2007.

347 For more information see the Department of Foreign Affairs (http://dfa.ie/home/index.aspx?id=3030), accessed 6 November 2007.

348 Bowyer KW (2004). Face Recognition Technology: Security versus Privacy. *IEEE Technology and Society Magazine* Spring 2004: 9–20.

349 Woodward JD Jr (2001). *Super Bowl Surveillance: Facing Up to Biometrics*. RAND, California, 16p.

350 Stanley J and Steinhardt B (2002). *Drawing a Blank: The failure of facial recognition technology in Tampa, Florida.* American Civil Liberties Union, New York, 9p. Available online at: http://www.aclu.org/FilesPDFs/drawing_blank.pdf, accessed 20 February 2008.

351 European Commission Joint Research Centre, Institute for Prospective Technology Studies (2005) *op. cit.*

352 Bowyer (2004) *op. cit.*

353 Woodward (2001) *op. cit.*

354 European Commission Joint Research Centre, Institute for Prospective Technology Studies (2005) *op. cit.*

photograph) after a specific time period to take account of such changes in appearance. The limited distinctiveness and the temporal variation in facial appearance can adversely affect the performance of this technology. System performance and accuracy is also influenced by differences in the lighting conditions (especially outdoors), the type of background, the distance to the sensor, the angle the images were captured at, the subject's expression, the visibility of the face (*e.g.* if occluded by hair, glasses or clothing), the level of cooperation of the subject/user, and the quality of the images.[355,356,357] These factors all influence the error rates (*i.e.* the FAR and the FRR) resulting in significant numbers of false positives and false negatives.[358] While incidences of poor performance in real world applications are well documented, the accuracy and overall performance of facial recognition systems in independent experimental tests is reported to be improving dramatically, for instance, for a FAR of 0.001 (1 in 1000), the corresponding FRR has dropped from 0.79 in 1993, to 0.01 in 2006.[359,360] Progress in sensor and algorithm development coupled with better control of lighting and environmental conditions (where possible) and guidance to the user can all help to improve performance further.[361,362]

The issues affecting the performance of facial recognition systems also contribute to their relatively low resistance to circumvention. Older systems and those that operate at a distance may be susceptible to disguises or deliberate non-cooperation of the subject, for instance, shielding the face from the camera. In addition, the higher FAR for facial recognition compared to some other modalities (*e.g.* iris and fingerprint) is also prone to attack. As noted above, implementing a more stringent matching threshold can reduce the FAR, but this results in an increased FRR and therefore greater inconvenience to legitimate users.

355 NSTC Subcommittee on Biometrics (2006c) *op. cit.*

356 Bowyer *et al.* (2006) *op. cit.*

357 Jain *et al.* (2004) *op. cit.*

358 For example, a recent trial at a rail terminal in Germany to identify people against a watch list gave between 30 and 60% correct matches [Biometric Technology Today (2007a) Face Recognition: Part One. *Biometric Technology Today* 15(9): 11]. A trial at Palm Beach International Airport in 2002 had a 47% successful match rate [Bowyer (2004) *op. cit.*], while in another trial at Logan Airport in Boston, also in 2002, the success rate was just over 50% [Marshall P (2007). We can see clearly now. *Government Computer News* June 4, 2007. Available online at: http://www.itl.nist.gov/iad/News/FaceRecog3.html, accessed 23 May 2008]. When employed at the Super Bowl in 2001, the facial recognition system gave zero correct matches [Marshall (2007) *op. cit.*].

359 Phillips PJ, Scruggs WT, O'Toole AJ, Flynn PJ, Bowyer KW, Schott CL and Sharpe M (2007). *FRVT 2006 and ICE 2006 Large-Scale Results.* Technical Report NISTIR 7408, National Institute of Standards and Technology, Maryland, 55p.

360 Biometric Technology Today (2007a) *op. cit.*

361 Phillips *et al.* (2007) *op. cit.*

362 Biometric Technology Today (2007a) *op. cit.*

Facial Thermography

Basic Information

Facial thermography measures the amount of thermal radiation (heat) emitted from an individual's face.[363,364,365] It has been suggested that the pattern of heat radiated by the human face (or body) is suitable for recognition purposes. An infrared camera is used to capture the heat images and analyse them for anatomical information, which is considered to be invariant to temperature changes, for example, the patterns of superficial blood vessels. The most likely applications of facial thermography are similar to those employing 2D- and 3D-based facial recognition. For example, this technology could be employed to secure computer and network access, at ATM cash dispensers and point of sale terminals and in e-passports. However, it has also been suggested that facial thermography recognition could potentially be used in the medical field for triage, diagnosis and monitoring treatments.

Critical Profile of Facial Thermography

All individuals produce facial thermograms and the complexity of blood vasculature underlying these thermograms is thought to be distinctive enough to permit recognition. However, facial thermograms can be affected by a number of different factors including the ambient temperature, the ingestion of certain substances (*e.g.* vasodilators and vasoconstrictors), extensive facial surgery, sinus problems, inflammation, arterial blockages, incipient stroke, soft tissue injuries and other physiological conditions. The collection of this biometric is unobtrusive, can be done at a distance and is possible under varying lighting conditions, including darkness. Despite this, difficulties can arise in capturing facial thermogram images in uncontrolled environments containing other heat sources.[366] Other factors can also reduce system performance such as the presence of glasses and even severe sunburn. Nonetheless, preliminary accuracy results seem promising and they are expected to improve.

As the facial thermograms are generated from blood vessels below the surface of the skin, this technology is resistant to circumvention using disguises. Moreover, attempts to change the pattern of blood vessels to alter the resulting thermogram can also be detected. Liveness detection is another possible security measure, for example, a number of image frames could be taken and analysed for small thermal variations caused by the heart rate and respiration. User acceptance of facial thermography is high owing to the non-contact, non-invasive nature of image collection and the fast throughput speed. Some concerns have been raised regarding the potential to infer certain medical conditions from the vascular patterns. In addition, because the images can be collected covertly, privacy concerns relating to surveillance have been raised.

---

363 NSTC Subcommittee on Biometrics (2006e) *op. cit.*

364 Prokoski FJ and Riedel RB (1999). Infrared Identification of Faces and Body Parts. In A Jain, R Bolle and S Pankanti (eds.) *Biometrics: Personal Identification in Networked Society*, Kluwer Press, Dordrecht, p.191–212.

365 This technology could potentially be used to recognise individuals, based on thermograms from other parts of the body.

366 Jain *et al.* (2004) *op. cit.*

Ear Geometry Biometrics

Basic Information

This form of biometric recognition is based on analyses of the shape of the outer ear, the ear lobes and bone structure,[367] and both 2D and 3D methodologies are used. A sensor (*e.g.* a camera) collects a side profile image of the user's head, from which the system automatically locates the ear and isolates it from the surrounding hair, regions of the face, and the user's clothes.[368] The algorithm uses a combination of colour and depth analysis to first localise the ear pit, then generates an outline of the visible ear region. The algorithm has to account for differences in skin tone (caused by lighting variation), as well as differences in ear size, ear shape, hair occlusion, and the presence of earrings.

Critical Profile of Ear Geometry Recognition

From a biometric perspective, ears present good universality and it has been suggested that the rich structure of the ear is unique enough to permit its use as a biometric.[369,370,371] However, others have questioned the level of distinctiveness of ear geometry, particularly for recognition purposes.[372] In the main, apart from injury, the structure of the ear is quite stable and undergoes only small, predictable changes over time, which can be accounted for in recognition systems. This is not the case, however, for very young individuals (*i.e.* 4 months to 8 years old) and the elderly (*i.e.* those over 70 years of age), for whom ear geometry exhibits more marked changes.[373,374]

Collectability is relatively straightforward, quick and non-invasive, which contributes to its high acceptability. In addition, while ear geometry can be collected passively, the overall performance is improved if the users are given feedback regarding their distance from the camera, their position and angle of exposure and their pose. Performance is also affected by a number of other factors including the occlusion of the ear by hair, clothing or earrings, and differences in illumination, which can increase specularity and shadowing of the ear structures. The 3D methodologies appear to cope better with some of these issues and preliminary results suggest that performance is improving. Overall, ear geometry recognition systems exhibit moderate resistance to circumvention.

---

367 OECD, Working Party on Information Security and Privacy (2004) *op. cit.*

368 For example, the algorithms may utilise texture and colour segmentation and/or thermal imaging to distinguish between the ear and hair in the images.

369 Yan P and Bowyer KW (2007). Biometric Recognition Using 3D Ear Shape. *IEEE Transactions on Pattern Analysis and Machine Intelligence* 29(8): 1297–1308.

370 Burge M and Burger W (1999). Ear Biometrics. In A Jain, R Bolle and S Pankanti (eds.) *Biometrics: Personal Identification in Networked Society*, Kluwer Press, Dordrecht, p.273–286.

371 While both a person's ears may be distinguishable from another's, the left and right ears of the same individual are approximately bilaterally symmetrical [Yan and Bowyer (2007) *op. cit.*].

372 Jain *et al.* (2004) *op. cit.*

373 Burge and Burger (1999) *op. cit.*

374 Yan and Bowyer (2007) *op. cit.*

Iris Recognition

The original concept of using the iris for recognition purposes was suggested in the 1930s, however, it was not until the early 1990s that an algorithm for automated iris recognition was developed.[375]

Basic Information

The iris is the coloured part of the eye around the pupil and is surrounded by the sclera (the white of the eye).[376,377] The purpose of biometric iris recognition is to enable a real time, high confidence confirmation of an individual's identity through the mathematical analysis of that individual's random iris patterns.[378] The user looks at the sensor, in this case a camera, and the detailed structure of his/her iris is illuminated using near infrared light (see Figures 7a and 7b).[379] The algorithm involved then produces a mathematical representation of the complex iris structure. The image is also modified to reduce noise and other irrelevant information caused by eyelashes and eyelids occluding (masking) the iris and to account for resolution issues due to the level of illumination.[380] This modification process can result in the loss of actual iris pattern information, but it is not considered to adversely affect the matching process. Finally, the remaining pixels relating to the iris are converted to bit pattern representations (templates or IrisCodes) of the iris, which are often up to 2048 bits in size.[381] During the recognition process a live iris image is converted to a template and is compared with the enrolled template *via* a bit-to-bit comparison, which measures the correlation between the irises.

375 National Science and Technology Council (NSTC) Subcommittee on Biometrics (2006i). *Iris Recognition*. NSTC, Washington, 10p. Available online at: http://www.biometrics.gov/Documents/IrisRec.pdf, accessed 10 October 2007.

376 Jain *et al.* (2004) *op. cit.*

377 OECD, Working Party on Information Security and Privacy (2004) *op. cit.*

378 Daugman J (2004). How Iris Recognition Works. *IEEE Transactions on Circuits and Systems for Video Technology* 14(1): 21–30.

379 Near infrared illumination is used because it reveals greater pattern complexity than visible light, especially for darkly pigmented irises [Daugman (2004) *op. cit.*]. However, the colour of the iris is not relevant for iris recognition since all scans produce black and white images.

380 Daugman (2004) *op. cit.*

381 *ibid.*

a)

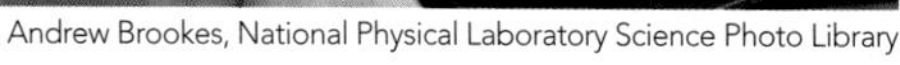

Andrew Brookes, National Physical Laboratory Science Photo Library

b)

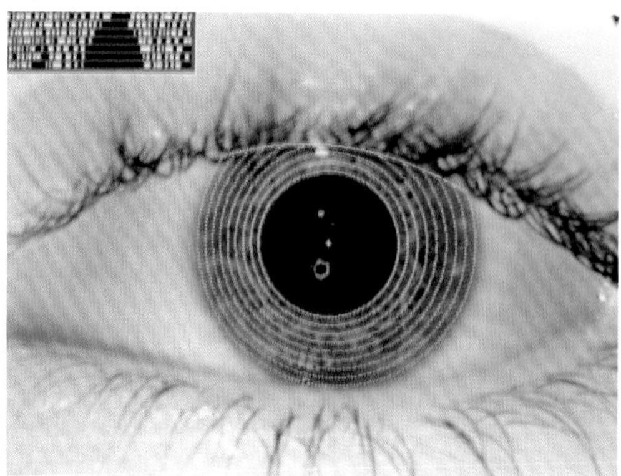

James King-Holmes/Science Photo Library

Figure 7: a) A man having his iris screened by a biometric scanner. The computer monitor is showing a match between this man and his profile held in the computer's records. b) A computer screen image of an iris being scanned. The computer converts certain iris features into a 256-byte code (shown inset). It divides the iris into 8 concentric circles (marked in white) so that landmarks can be recognised even when they are compressed when the pupil is dilated.[382]

Applications of Iris Recognition

Iris recognition systems operate well in both verification- and identification-based applications. For example, the UK government has implemented such a system for immigration purposes, *i.e.* the Iris Recognition Immigration System (IRIS). This system offers a fast and secure way to pass through immigration. Once an individual has enrolled in the system he/she just needs to scan his/her iris at the appropriate sensors and if it matches the template in the database he/she can pass through border control. It is envisaged that, once enrolled, passage through the IRIS barrier should take approximately 20 seconds.[383] Currently, IRIS barriers are in place at all Heathrow terminals, Gatwick North and South, Manchester Terminals 1 and 2 and Birmingham Airport. A similar system is already in place in Amsterdam's Schiphol Airport and a number of US and Canadian airports.

382 The IriScan system shown in Figure 7b was developed by Dr John Daugman of Cambridge University, Britain.

383 UK Home Office, Border and Immigration Agency.
Available online at: http://www.ind.homeoffice.gov.uk/managingborders/technology/iris/

In the United Arab Emirates (UAE) an iris recognition system is also used for border control purposes. All foreign nationals arriving at any international airport, land port or seaport have their irises screened against a database of individuals who have been expelled from the UAE.[384] It is estimated that this system screens approximately 12,000 passengers each day and the system has resulted in over 73,000 matches, with no false matches reported.[385]

Critical Profile of Iris Recognition

Irises exhibit high universality, since almost all humans possess irises. However, there are limited exceptions of people who possess no irises, either for genetic reasons (*i.e.* aniridia[386]) or as a result of medical intervention, for example, laser iridotomy, used to correct glaucoma.[387] The visual texture or pattern of the iris is considered an epigenetic trait,[388] which is formed during the development of the foetus and stabilises during the first two years of life. As a result, the iris texture is considered unique, even between both eyes of the same individual or the eyes of identical twins. The position of the iris inside the eye offers it some degree of protection and, consequently, iris patterns are predominantly stable over time, though they can be affected by certain eye diseases in a minority of individuals.

The collectability of iris images can be problematic at times. User cooperation is essential and additional feedback or training may also be required to ensure the user's head and eyes are positioned correctly and at an appropriate distance from the camera. As a result, enrolment can take some time (from 30 seconds to 2–3 minutes),[389] though it should be noted that for accustomed users, image collection is extremely quick.[390] The user is usually positioned 10–20 cm from the camera, though research is assessing the ability to record iris patterns at distances of up to 3–5 m. There have even been developments in the ability to perform iris recognition on the move.[391] Iris recognition systems are considered to be among the most accurate of biometric systems, even in large-scale, identification-based applications and accuracy is constantly improving. For example, in tests conducted by the UK government in 2001, the algorithms involved produced no false matches and only 0.2 per cent false rejections in over 2.3 million iris comparison tests.[392] More recently, a number of third-party test studies reported

384 Daugman J and Malhas I (2004) Iris recognition border-crossing system in the UAE. *International Airport Review* 2: 49–53.

385 Daugman J (2008) *United Arab Emirates Deployment of Iris Recognition.* Available online at: http://www.cl.cam.ac.uk/~jgd1000/deployments.html, accessed 15 February 2008.

386 Aniridia is estimated to affect 1.8 in every 100,000 births. [European Commission Joint Research Centre, Institute for Prospective Technology Studies (2005) *op. cit.*].

387 European Commission Joint Research Centre, Institute for Prospective Technology Studies (2005) *op. cit.*

388 Epigenetic: Refers to heritable factors affecting the development or function of an organism that are not associated with its DNA sequence. Available online at: www.everythingbio.com, accessed 12 June 2008.

389 European Commission Joint Research Centre, Institute for Prospective Technology Studies (2005) *op. cit.*

390 Iris recognition processing can be done in a matter of seconds.

391 Matey JR, Naroditsky O, Hanna K, Kolczynski R, LoIacono DJ, Mangru S, *et al.* (2006). Iris on the Move: Acquisition of Images for Iris Recognition in Less Constrained Environments. *Proceedings of the IEEE* 94(11): 1936–1947.

392 Daugman and Malhas (2004) *op. cit.*

false non-match rates ranging from 0.01 to 0.04 at a false match rate of 0.001 (one in 1000).[393] However, performance can be affected by certain image acquisition conditions – occlusion of the iris by the eyelids and eyelashes, for example – which may be substantial enough to require re-enrolment or recollection of the image. Differences in illumination and the angle of image capture between enrolment and subsequent sampling episodes can also cause problems, although newer sensors appear to be less susceptible to these issues. In addition, the cameras used can usually account for the presence of glasses or contact lenses and the feature extraction algorithms can generally cope with other variations, for example, in iris size, the distance from the camera, the magnification and the degree of pupil dilation.[394] Iris recognition systems are also difficult to circumvent, especially newer systems, which are less susceptible to spoofing using fake irises. Efforts of liveness detection, such as assessing variations in pupil dilation have also improved and human supervision can reduce the opportunities for spoofing. Furthermore, if a template (IrisCode) is compromised, it is possible to generate a new one for that individual.[395]

Iris recognition systems experience relatively low acceptability, although levels of acceptance are improving.[396] The technology is often considered invasive and there are common misconceptions that this technology uses lasers (which it does not) and that there is a risk of potential damage to the eye. There is, currently, no evidence to suggest that the near infrared illumination used by the sensors causes any damage to the eyes.[397,398] Advances in the ability to capture iris images at a distance and on the move have heightened concerns regarding the potential tracking and surveillance of individuals. However, while initial enrolment times can be relatively slow, the speed of processing for accustomed users makes it very convenient.

As it stands, the diffusion of iris recognition systems has been hindered by the high costs of the sensors, the potential for system operators being confined to a particular algorithm or technology and the perception of discomfort to the user. However, sensor costs are decreasing and user acceptance is increasing and it is anticipated that the deployment of iris recognition systems will increase in the coming years.[399]

---

393 Newton EM and Phillips PJ (2007) *Meta-Analysis of Third-Party Evaluations of Iris Recognition*. Technical Report NISTIR 7440, National Institute of Standards and Technology, Maryland, 14p.

394 Daugman (2004) *op. cit.*

395 European Commission Joint Research Centre, Institute for Prospective Technology Studies (2005) *op. cit.*

396 *ibid.*

397 *ibid.*

398 Matey *et al.* (2006) *op. cit.*

399 European Commission Joint Research Centre, Institute for Prospective Technology Studies (2005) *op. cit.*

Retina

Basic Information

Biometric retina recognition is based on the comparison of the complex pattern of blood vessels located at the back of the eye.[400,401] While research in this field has been ongoing since the 1970s, retina recognition is not currently a major player in the biometrics market. The user looks into an eyepiece and focuses on a designated point in the viewing field, which helps align the eye correctly and fixes the area of the retina that will be imaged.[402,403] Near infrared light, which is invisible to the user, illuminates the vascular network of the retina and this is reflected back to the sensor as a wavelength. The algorithm then creates a unique "signature" (template) based on the blood vessel pattern of the retina of that individual.[404] Owing to the small size of the area to be imaged, the user needs to be quite close to the sensor during image capture, usually between approximately 2 and 5 cm away.[405] However, research is being conducted on sensors that can capture images from a distance of approximately 30 cm.[406] Retina recognition systems are expensive and tend to have low acceptance levels and they are not widely utilised outside high security and national security applications.[407]

Critical Profile of Retina Recognition

Similarly to iris patterns, retina vasculature is thought to be unique for each individual and for both eyes of the same individual.[408] The position of the retina at the back of the eye keeps it well protected, and, consequently, the vascular network exhibits very high stability over time.[409] While the universality of the retina is high in the general population, the collectability of this biometric can be problematic.[410,411] For example, multiple areas of the retina could potentially be presented to the sensor, therefore each user needs to be trained in using the equipment to align his/her eye correctly. This requires the cooperation of the user as well as his/her conscious effort to keep his/her head still and focus on the alignment point/light in the sensor, which can take some time.[412] Despite this training, individuals with certain medical conditions (*e.g.* astigmatism) may still encounter problems in aligning their eyes correctly to facilitate scanning.

---

400 Woodward *et al.* (2001) *op. cit.*

401 NSTC Subcommittee on Biometrics (2006c) *op. cit.*

402 Jain *et al.* (2004) *op. cit.*

403 Hill R (1999). Retina Identification. In A Jain, R Bolle and S Pankanti (eds.) *Biometrics: Personal Identification in Networked Society*, Kluwer Press, Dordrecht, p.123–142.

404 *ibid.*

405 Woodward *et al.* (2001) *op. cit.*

406 Hill (1999) *op. cit.*

407 OECD, Working Party on Information Security and Privacy (2004) *op. cit.*

408 Jain *et al.* (2004) *op. cit.*

409 Hill (1999) *op. cit.*

410 OECD, Working Party on Information Security and Privacy (2004) *op. cit.*

411 Jain *et al.* (2004) *op. cit.*

412 It has been shown that following successful enrolment, image capture times can be quite quick, *e.g.* 6–10 seconds [OECD, Working Party on Information Security and Privacy (2004) *op. cit.*].

Once the user becomes accustomed to using the scanning equipment, retinal recognition offers a highly accurate performance in both verification and identification modes. Notwithstanding the high performance of this biometric technology, it is not widely accepted. Scanning of the retina is often considered to be invasive and health concerns have been raised relating to potential thermal damage to the eye. While some of these concerns can be assuaged by clarifying that the scanning process does not involve a laser, the potential for damage still needs to be ascertained.[413] Images of an individual's retinal vasculature can reveal additional medical information about that person, for example, high blood pressure, pregnancy and AIDS, which raises serious privacy concerns.[414] Finally, retina recognition is considered to be one of the most secure biometrics as it is difficult to change or replicate the retinal vasculature.[415]

Gait

Basic Information

Gait is a complicated spatio-temporal biometric, which relates to the specific way an individual walks.[416,417] Moreover, humans have been shown to identify and recognise people on the basis of their gait.[418] In terms of biometric recognition of gait, a video camera is used to capture the specific repeating pattern produced by an individual as he/she walks. An algorithm is used to determine the mathematical relationship between each point of movement of the body and to create a signature pattern (template) necessary for recognition.[419,420] Biometric gait recognition can utilise the shape and/or the dynamics of the body as it moves and these are predominantly assessed through silhouette matching.[421] Other factors such as stride length, cadence and stride speed as well as static body movements may also be assessed.

Applications of Gait Recognition

Gait recognition technology could potentially be used for video surveillance and security purposes.[422] Other potential uses of gait recognition are being investigated as part of the EU-based ACTIBIO project (Unobtrusive Authentication Using ACTIvity Related and Soft BIOmetrics). This project combines a number of different biometric

413 European Commission Joint Research Centre, Institute for Prospective Technology Studies (2005) *op. cit.*

414 Woodward *et al.* (2001) *op. cit.*

415 Jain *et al.* (2004) *op. cit.*

416 Nixon MS, Carter JN, Cunado D, Huang PS and Stevenage SV (1999). Automatic Gait Recognition. In A Jain, R Bolle and S Pankanti (eds.) *Biometrics: Personal Identification in Networked Society*, Kluwer Press, Dordrecht, p.231–250.

417 Jain *et al.* (2004) *op. cit.*

418 Sarkar S, Phillips PJ, Liu Z, Vega IR, Grother P and Bowyer KW (2005). The HumanID Gait Challenge Problem: Data Sets, Performance, and Analysis. *IEEE Transactions on Pattern Analysis and Machine Intelligence* 27(2): 162–177.

419 Sarkar S and Liu Z (2008). Gait Recognition. In AK Jain, P Flynn and AA Ross (eds.) *Handbook of Biometrics*, Springer, New York, p.109–129.

420 OECD, Working Party on Information Security and Privacy (2004) *op. cit.*

421 Sarkar and Liu (2008) *op. cit.*

422 Sarkar *et al.* (2005) *op. cit.*

modalities, including gait,[423] and utilises unobtrusive sensors[424] to develop adaptable, dependable and secure recognition systems. The initial research will focus on three separate scenarios: (i) a security operator system for indoor premises; (ii) a continuous authentication system for vehicle drivers to prevent hijacking; and (iii) authentication *via* activity recognition and control in transactions through "always-on machines", for example, in an office workspace.[425]

Critical Profile of Gait Recognition

Gait is not a universal biometric trait, since not all individuals are able to walk. In addition, gait is not regarded to be very distinctive across large populations, but it is considered sufficiently distinguishable for recognition purposes in low security applications.[426] While an individual's gait is influenced by his/her musculo-skeletal structure it is a behavioural trait, and is prone to variation over time, for example, due to changes in body weight, pregnancy, injuries (especially to the legs or feet) and even drunkenness.[427,428]

An individual's gait can be collected from a distance and from a number of angles, even using a low resolution video camera. Collection can also be achieved with or without the user's cooperation or knowledge. While the ability to examine an individual's gait covertly and at a distance may raise some concerns relating to surveillance, gait recognition systems are generally widely accepted. However, this type of system is not considered to offer very high performance overall. While indoor applications of gait recognition have shown somewhat better performance levels, this technology is most likely to be employed outdoors, where changing environmental conditions, for example, illumination and the presence of shadows, can affect recognition accuracy adversely.[429] Furthermore, it has been shown that performance is also particularly susceptible to differences in footwear, clothing, walking surface, walking speed, whether or not the individual was carrying something, whether image collection occurred indoors or outdoors, and the time elapsed since the individual last used the system.[430] Research is ongoing to try to overcome some of these issues and also to try and identify aspects of gait that are not affected by these factors. While it has been suggested that it could be difficult for an individual to mask his/her own gait pattern while posing as someone else, the current level of performance of these systems would leave it open to circumvention.

---

423 The other modalities include face, gesture, body dynamics and analysis of the user's EEG (Electroencephalogram) and ECG (Electrocardiogram).

424 The research is examining wearable sensors and other integrated sensors (*e.g.* into seats, or sound-based recognition sensors).

425 For more information see the ACTIBIO project website: http://www.actibio.eu:8080/actibio/index.html, accessed 28 February 2008.

426 Jain *et al.* (2004) *op. cit.*

427 Nixon *et al.* (1999) *op. cit.*

428 Jain *et al.* (2004) *op. cit.*

429 Sarkar *et al.* (2005) *op. cit.*

430 Sarkar and Liu (2008) *op. cit.*

Odour Recognition

Basic Information

These systems are based on the recognition of characteristic components of odour emitted by a given individual.[431] Since odour is emitted from pores all over an individual's body, these systems operate by circulating air around the body part being analysed (*e.g.* the back of the hand, the arm or the neck) and over an array of chemical sensors. Each of these sensors is sensitive and receptive to certain groups of aromatic compounds of the individual's smell,[432] which are extracted and classified into a template.[433]

Critical Profile of Odour Recognition

All individuals emit an odour, components of which are considered to be distinctive. While the odour profile itself is considered permanent, it can be affected by certain foods and medications.[434] While body odour can be collected from non-intrusive parts of the body, currently available sensors have difficulties in distinguishing the invariant components of body odour, which limits system performance. In addition, performance can also be affected by the use of deodorants and perfumes[435] and contamination or odour transfer between different people.[436]

Voice Recognition

Basic Information

Voice is often classified as a combination of a behavioural and a physiological biometric because certain features of an individual's voice are based on the shape and size of their vocal tracts, mouth, nasal cavities, lips, *etc.*[437,438,439] From a biometrics perspective there are basically two different types of voice/speaker recognition system, *i.e.* text dependent and text independent systems.[440,441] In a text dependent system the user speaks a particular, predetermined, pass phrase, for example, a sequence of numbers. When enrolling in such a system, the user may be required to repeat the pass phrase a number of times,[442] to enable the algorithm to take account of any intra-class variation. Consequently, the enrolment process lasts longer, but this is thought to result in increased accuracy.

---

431 Jain *et al.* (2004) *op. cit.*

432 *ibid.*

433 OECD, Working Party on Information Security and Privacy (2004) *op. cit.*

434 Persaud KC, Lee D-H and Byun H-G (1999). Objective Odour Measurements. In A Jain, R Bolle and S Pankanti (eds.) *Biometrics: Personal Identification in Networked Society*, Kluwer Press, Dordrecht, p.251–272.

435 Heyer R (2008). *Biometric Technology Review 2008.* Land Operations Division, (DSTO) Defence Science and Technology Organisation, Australia, 60p.

436 Persaud *et al.* (1999) *op. cit.*

437 Jain *et al.* (2004) *op. cit.*

438 National Science and Technology Council (NSTC) Subcommittee on Biometrics (2006j). *Speaker Recognition.* NSTC, Washington, 9p. Available online at: http://www.biometrics.gov/Documents/SpeakerRec.pdf, accessed 10 October 2007.

439 Campbell JP (1999). Speaker Recognition. In A Jain, R Bolle and S Pankanti (eds.) *Biometrics: Personal Identification in Networked Society*, Kluwer Press, Dordrecht, p.165–190.

440 Jain *et al.* (2004) *op. cit.*

441 Heyer (2008) *op. cit.*

442 OECD, Working Party on Information Security and Privacy (2004) *op. cit.*

In a text independent system the user's voice is recognised regardless of what he/she is saying. Such systems are said to offer greater security against abuse than text dependent systems, but they are more difficult to design.[443] In general, sound waves from the individual's voice recording are calculated as feature vectors, which are then modelled as a voiceprint (template) for that individual (see Figure 8).[444,445] During the recognition process, the sequences of feature vectors from the sample and enrolled voiceprints are compared using pattern analysis, *i.e.* the system *does not* compare the voice itself. If these patterns are sufficiently similar, a match is given.

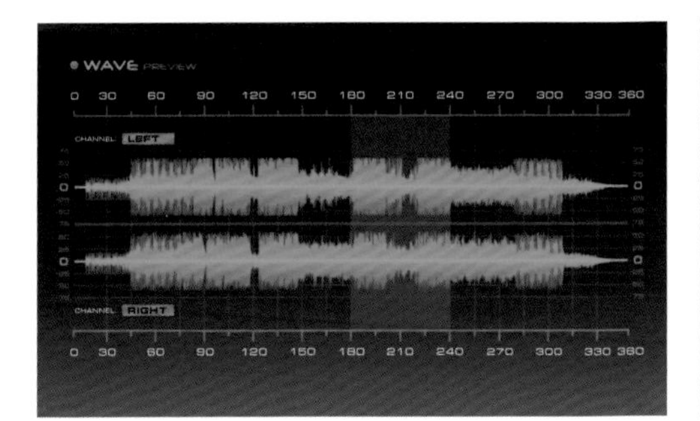

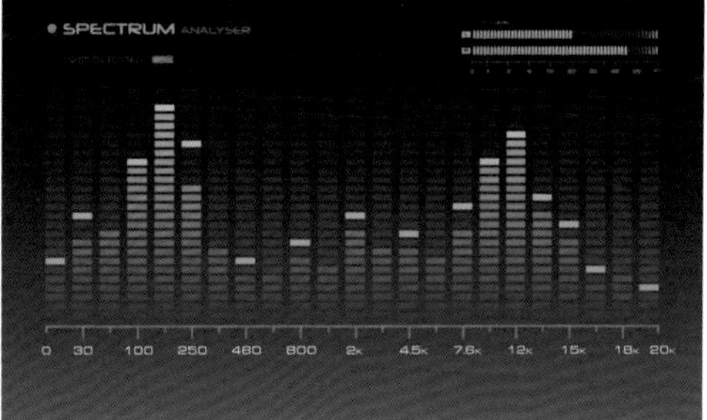

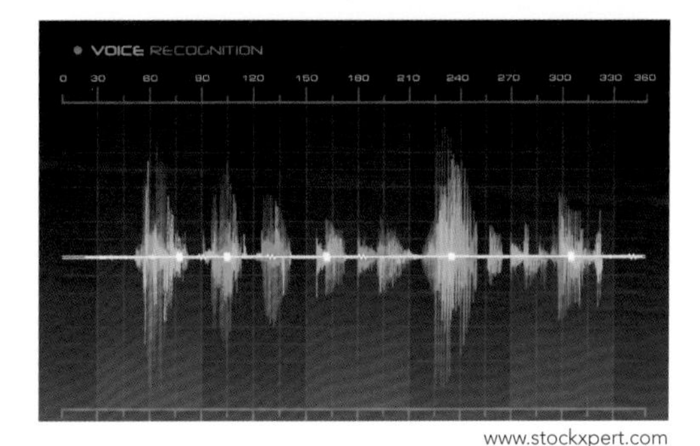

www.stockxpert.com

Figure 8: An example of recordings for a given individual from a voice recognition programme.

443 Jain *et al.* (2004) *op. cit.*

444 Campbell (1999) *op. cit.*

445 NSTC Subcommittee on Biometrics (2006j) *op. cit.*

Applications of Voice Recognition

Voice biometrics are usually used in verification-based applications, and have been implemented in the financial services sector, especially e-commerce and e-banking.[446] For example, telephone banking in a number of countries (*e.g.* the US, Brazil and Israel) uses voice recognition to enable customers to access their accounts, conduct transactions and change passwords and PINs.[447] Voice recognition is also developing as a means of paying for goods and services *via* the telephone, for instance, through companies such as VoicePay.[448] In this case an individual opens an account with VoicePay over the telephone (*i.e.* they provide a username, a password and details of a payment card), and VoicePay produces a voiceprint from that individual's voice recording. Then when purchasing a product or service over the Internet or in a shop the individual would be required to repeat two randomly generated number codes over the telephone for comparison with his/her voiceprint. If the system returns a match the transaction can be completed.[449,450,451]

From an Irish perspective, Buywayz Ltd[452] utilises voice recognition systems to enable farmers to conduct business transactions over the telephone.[453] Each individual farmer registers as a member of Buywayz, which involves providing relevant details relating to him/her (*e.g.* name, registered phone number and banking/payment details) and Buywayz takes a voice sample to generate their voiceprint. When an individual member wishes to place an order he/she calls the Buywayz system, provides a voice sample, which is compared with the stored voiceprint to verify his/her identity. Following verification, the individual keys in the product order number (obtained from the Buywayz website or *via* a text message) and the quantity he/she wishes to purchase into his/her telephone, the system then confirms this information and sends it to the supplier *via* the offer management system.

Voice recognition is also used in law enforcement for forensic purposes, whereby voice recordings of an individual taken when in police custody are compared with legally intercepted conversations collected as part of an investigation. If the system gives a match, this can be corroborated by a forensic expert and used in court as evidence,[454,455] which has happened in a number of countries, including Spain, Germany, France,

---

446 The Economist (2006) *op. cit.*

447 Most (2004a) *op. cit.*

448 For more information see the VoicePay website available online at: http://www.voice-pay.com, accessed 6 August 2009.

449 The Economist (2007). Mobile commerce. A better way to pay by phone. *The Economist* 19 July 2007. Available online at: http://www.economist.com/science/PrinterFriendly.cfm?story_id=9507446, accessed 17 October 2007.

450 Ogden N (2007). *VoicePay – so let's talk money.* Presentation at the Biometrics Exhibition and Conference 2007, 17–19 October 2007, Westminster, London.

451 However, it should be noted that the VoicePay system is not, as of yet, universally available and only products that are registered with VoicePay can be purchased in this way.

452 Buywayz is a limited company established as a joint venture between The Irish Farmer's Association, VoiceVault Ltd. and IMS MAXIMS plc.

453 For more information see the Buywayz Limited website, available online at: www.buywayz.com, accessed 12 June 2008.

454 Martinez E (2007). *Case History: Spanish police implement world's first automatic speaker identification system (ASIS).* Presentation at the Biometrics Exhibition and Conference 2007, 17–19 October 2007, Westminster, London.

455 For more information see the Agnitio website, available online at: http://www.agnitio.es/ingles/asis.php, accessed 23 May 2008.

Columbia, Chile and South Korea.[456] In addition, a database of voice recordings from criminals can be built up for use in Automatic Speaker Identification Systems (ASISs), such as the SAIVOX (Sistema Automático de Identificación por Voz) used by the Spanish police. This system is used in the investigation of crimes that routinely involve intercepting telephone conversations (*e.g.* terrorism, drugs and organised crime). Intercepted recordings of unknown speakers are compared with all the recordings in the SAIVOX database, which may result in a match with a known criminal already in the database. Alternatively, a newly captured voice recording could match an older recording from an unknown suspect, thus helping to link different ongoing investigations.[457]

Critical Profile of Voice Recognition

Voice exhibits a relatively high level of universality since most people can speak; however, there are some exceptions such as those who temporarily (or permanently) lose their voices or those people who where never able to speak. The degree of distinctiveness of individual voices is, generally, not considered sufficient for large-scale databases.[458,459] Overall, voice biometrics are regarded as moderately stable over time, with the physiological characteristics of an individual's voice usually invariant (barring injury or possibly surgery). However, the behavioural component of an individual's voice is prone to change over time due to ageing, medical conditions (*e.g.* if the user has a cold) and emotional state.[460,461] Voice recognition systems, therefore, tend to utilise the physiological voice components more often.

Voice biometrics are quick and easy to collect, even remotely, requiring only a simple audio capture device, such as a telephone. In fact, these systems can be linked to existing telephone lines and/or computer networks, which facilitates the establishment of such recognition systems. As noted above, the familiarity with this form of recognition, allied to the ease of use and non-invasive nature of this technology have resulted in its high level of acceptability.[462,463] Notwithstanding, the general acceptance of this technology, problems with the telephone line (or channel) quality and background noise can adversely affect the performance of these systems.[464] Differences in the voice capturing device used during enrolment and that used for subsequent recognition also impinges on performance.[465] The accuracy and reliability of newer voice recognition technologies are improving as these systems are reputed to be more robust to noise, channel variation and human changes, as well as mimicry by humans and tape

456 Martinez (2007) *op. cit.*
457 *ibid.*
458 Jain *et al.* (2004) *op. cit.*
459 OECD, Working Party on Information Security and Privacy (2004) *op. cit.*
460 Jain *et al.* (2004) *op. cit.*
461 Dessimoz *et al.* (2006) *op. cit.*
462 Huijgens (2007) *op. cit.*
463 OECD, Working Party on Information Security and Privacy (2004) *op. cit.*
464 Dessimoz *et al.* (2006) *op. cit.*
465 NSTC Subcommittee on Biometrics (2006i) *op. cit.*

recorders. It is envisaged that system accuracy will continue to increase in the future, although it has been suggested that currently these systems are not suited to large-scale identification applications.[466] Advances in this technology have improved system resistance to mimicry and tape recordings, for example, through the use of interaction-based liveness detection and authentication.

Keystroke Dynamics

Basic Information

It has been suggested that individuals have a characteristic way of typing on a keyboard, which is sufficient for use in biometric recognition systems.[467,468] This technology can assess an individual's keystroke dynamics (*e.g.* speed and pressure), the total typing time for a specific password and the time taken between hitting certain keys.[469,470] Keystroke dynamic systems are moderately resistant to circumvention, but they are usually used for low security applications, *e.g.* for controlling and monitoring access to computer systems and networks.[471]

Critical Profile of Keystroke Dynamics

While not considered unique to a given individual, it has been suggested that keystroke dynamics are distinctive enough to verify an individual's identity.[472] However, not all individuals can exhibit keystroke dynamics, for example, due to insufficient literacy levels or competence in using computers. As a behavioural biometric trait, keystroke dynamics are inherently variable over time. This variation, combined with the limited distinctiveness of this biometric, results in poor system performance, which limits the implementation of this technology to small-scale applications. This trait can be collected quite easily and unobtrusively,[473] which may assist in the acceptance of this method of recognition.

---

466 Jain *et al.* (2004) *op. cit.*

467 Obaidat MS and Sadoun B (1999). Keystroke Dynamics Based Authentication. In A Jain, R Bolle and S Pankanti (eds.) *Biometrics: Personal Identification in Networked Society*, Kluwer Press, Dordrecht, p.213–230.

468 Jain *et al.* (2004) *op. cit.*

469 Woodward *et al.* (2001) *op. cit.*

470 Obaidat and Sadoun (1999) *op. cit.*

471 Woodward *et al.* (2001) *op. cit.*

472 It should be noted that not all keystrokes characterise a specific key pattern.

473 Heyer (2008) *op. cit.*

Dynamic Signature

Basic Information

The way in which an individual signs his/her name is considered to be characteristic of that person and as such could provide a feasible mode of biometric recognition. Dynamic signature recognition is an automated method of examining an individual's signature.[474,475] These systems assess specific features of the signature writing process, including the speed, direction and pressure of writing, the time the stylus (*e.g.* a pen) is in and out of contact with the surface (*e.g.* paper), the total time taken to write the signature and where the stylus is raised and lowered on the surface.[476,477] It has been suggested that automated signature recognition should measure the degree of similarity in signature shapes.[478] In addition, during enrolment the signature should be collected a number of times to provide a more representative indication of intra-class variation. Moreover, research in this area should try to develop signature models and algorithms that are better able to adapt to and cope with such intra-class variation.[479] Dynamic signature recognition systems are used to authenticate electronic documents in hospitals, pharmacies and insurance firms in the US.[480]

Critical Profile of Dynamic Signature Recognition

Signature is not a universal biometric because a large proportion of individuals are unable to write owing to illiteracy problems and even certain medical conditions and injuries. While signatures are not considered to be very distinctive, they have been accepted as a means of verification for various government, legal, financial and commercial transactions.[481] Since signature dynamics is a behavioural biometric trait it is liable to change over time,[482,483] for example, it can be affected by an individual's physical or emotional state. As a result, signature dynamics is not envisaged as a very robust and stable biometric characteristic. The acceptability of this biometric is facilitated by its ease of collection and the familiarity of using ordinary written signatures for verification. Nonetheless, the performance of signature dynamics for biometric recognition is not considered to be very accurate.[484] Moreover, while there is, reportedly, less scope to forge signature dynamics as opposed to forging a signature itself, system security can be further enhanced through the comparison of both temporal and dynamic features of the signatures as well as their shapes.

---

474 Woodward *et al.* (2001) *op. cit.*

475 Normal, "offline" signature verification involves checking that the sample signature (*e.g.* on a transaction receipt) resembles the stored signature (*e.g.* on a credit card). This process can be quite subjective.

476 Woodward *et al.* (2001) *op. cit.*

477 Nalwa VS (1999). Automatic On-line Signature Verification. In A Jain, R Bolle and S Pankanti (eds.) *Biometrics: Personal Identification in Networked Society*, Kluwer Press, Dordrecht, p.143–164.

478 *ibid.*

479 *ibid.*

480 OECD, Working Party on Information Security and Privacy (2004) *op. cit.*

481 Jain *et al.* (2004) *op. cit.*

482 OECD, Working Party on Information Security and Privacy (2004) *op. cit.*

483 National Science and Technology Council (NSTC) Subcommittee on Biometrics (2006k). *Dynamic Signature.* NSTC, Washington, 7p. Available online at: http://www.biometrics.gov/Documents/DynamicSig.pdf, accessed 10 October 2007.

484 OECD, Working Party on Information Security and Privacy (2004) *op. cit.*

DNA (Deoxyribonucleic Acid)

Basic Information

Each individual human is identifiable by genetic variation found in his/her DNA, which is contained in the nucleus of almost every cell as well as mitochondria.[485] DNA serves as a unique genetic code, half of which comes from each parent. Identical twins are the exception to this rule since they have the same genetic code.[486] DNA is a long double stranded molecule that is composed of four bases: (i) adenine, (ii) guanine, (iii) cytosine and (iv) thymine. In the case of humans, there are approximately three billion bases, 99 per cent of which are the same from person to person. The variations, or order of the bases, in the remaining 1 per cent are the means by which DNA becomes unique to each individual. This remaining 1 per cent can be used to identify or verify the identification of a given individual. As there are so many bases in a person's DNA, the task of analysing all of them would be impracticable, thus, scientists use a small number of sequences of DNA (short tandem repeats) that are know to vary greatly among individuals to ascertain identity.

Despite the fact that DNA profiling is recognised as the most consistently effective method of establishing a permanent record of identity (statistical sampling shows a one in six billion chance of two people having the same profile), its role as a method of identity verification currently remains limited. This is largely because the process of producing a DNA profile is not automatic and cannot be conducted in real time, *i.e.* it takes a few hours.[487,488] Moreover, unlike other biometrics, DNA profiling requires the removal of material from the body itself rather than feature extraction or template generation and this inevitably raises issues in relation to bodily integrity. Thus, the differences between traditional biometrics and DNA are at this point in time distinct and make a full discussion of DNA as a biometric identifier outside the scope of the current report. Nonetheless, the level of accuracy of DNA, as indicated by its use in forensic applications (*e.g.* for law enforcement) and for paternity testing, suggest that it could potentially be used for biometric recognition in the future and therefore merits a limited discussion.

Forensic DNA identification is based on the process of DNA profiling. This involves the analysis of the numbers of tandemly repeating sequences of non-coding DNA, *i.e.* regions of DNA that are not part of genes and are, generally, not considered to have any specific function, from a given locus on the human genome.[489,490,491] Depending on the exact methodology used, a particular number of loci may be targeted, which are from different parts of the DNA

485 Rudin N, Inman K, Stolovitzky G and Rigoutsos I (1999). DNA Based Identification. In A Jain, R Bolle and S Pankanti (eds.) *Biometrics: Personal Identification in Networked Society*, Kluwer Press, Dordrecht, p.287–310.

486 Law Reform Commission (2005). *Report – The Establishment of a DNA Database.* (LRC 78-2005), Law Reform Commission, Dublin, 129p.

487 Jain *et al.* (2004) *op. cit.*

488 European Commission Joint Research Centre, Institute for Prospective Technology Studies (2005) *op. cit.*

489 Law Reform Commission (2005) *op. cit.*

490 Nuffield Council on Bioethics (2007). *The Forensic Use of Bioinformation: ethical issues.* Nuffield Council on Bioethics, London, 139p. Available online at: http://www.nuffieldbioethics.org/fileLibrary/pdf/The_forensic_use_of_bioinformation_-_ethical_issues.pdf, accessed 31 October 2007.

491 Each piece of DNA is made up of two strands, *i.e.* one from each parent, and these may contain different numbers of the repeats fragments at a given locus.

molecule.[492] In the case of forensics, the first stage in DNA analysis involves the collection of a DNA sample (a collection of cells), such as blood or hair,[493] for example, from a crime scene. The DNA is then isolated from this sample and the targeted loci are first amplified,[494] then the DNA is cut and sorted so that the different sections are arranged by size, *i.e.* related to the number of repeating units.[495,496] The final DNA profile when transcribed is a digital representation of the requisite areas of variability with the number of repeat units at each locus indicated (see Figure 9).[497]

With forensic DNA identification, two DNA profiles, for example, one taken from the scene of a crime and a reference profile generated from a criminal suspect[498] are compared. If both DNA profiles are different, the individual suspect is unlikely to be the source of the sample from the crime scene.[499,500] If the DNA profiles match, then the question arises whether or not the DNA sample collected from crime scene is actually from the suspect or from someone else with the same DNA profile. The significance of the match is dependent on the number of loci that are compared, for example, the probability of two profiles from two different people matching exactly over ten or more loci is considered to be one in one billion (except in the case of identical twins).[501,502] The result is related directly to the frequency of a particular allele in the population; therefore, if multiple alleles between two DNA profiles match, this increases the likelihood that they came from the same individual.[503] Despite this, a number of factors can increase the likelihood of a false match occurring. For example, if the original sample contained only a small amount of DNA; if only a small number of loci are compared; if the DNA profile from the crime scene is incomplete or degraded in some way; or if either DNA profile was contaminated. Therefore, it should be noted that DNA profiling is not foolproof.

492 Law Reform Commission (2005) *op. cit.*

493 Apart from blood and hair, other body materials including urine, semen, saliva, skin, teeth and bone can provide a DNA sample.

494 Small samples of DNA can be amplified in the laboratory, through a process known as a polymerase chain reaction (PCR), which produces multiple copies of the DNA.

495 Law Reform Commission (2005) *op. cit.*

496 European Commission Joint Research Centre, Institute for Prospective Technology Studies (2005) *op. cit.*

497 Law Reform Commission (2005) *op. cit.*

498 Profiles from other non-suspect individuals may also be compared with the crime scene profile during criminal investigations.

499 Rudin *et al.* (1999) *op. cit.*

500 European Commission Joint Research Centre, Institute for Prospective Technology Studies (2005) *op. cit.*

501 Nuffield Council on Bioethics (2007) *op. cit.*

502 European Commission Joint Research Centre, Institute for Prospective Technology Studies (2005) *op. cit.*

503 Law Reform Commission (2005) *op. cit.*

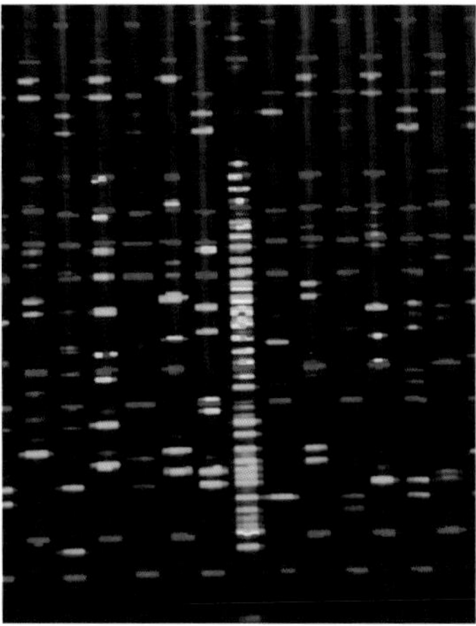

Tek Image/Science Photo Library

Figure 9. A DNA profile on a computer monitor screen. Each coloured band represents one of the bases that make up the genetic code of this sample of DNA.

Applications of DNA Recognition

DNA-based identification systems have limited commercial uses and this technology is mostly used for paternity tests, criminal identification and forensics.[504] Evidence based on DNA identification is routinely accepted and increasingly demanded in judicial proceedings. Many countries have established DNA banks for the purpose of collecting and storing DNA samples/information of suspects and those convicted of crimes. The practice of DNA profiling now forms an integral part of criminal investigations and Ireland is currently committed to the establishment of a forensic DNA database. In 2005, the Law Reform Commission published a comprehensive report on the constitutional and human rights issues association with the establishment of a forensic DNA database.[505]

There has been discussion in several countries about the possibility in the near future of incorporating DNA profiles, stored on an electronic chip, into identity documentation such as ID cards and passports. Machine readable DNA profiles could be checked against an existing archive of records, thus extending the use of DNA databases beyond current forensic applications. In 2007, the French parliament passed a highly controversial bill, which introduced DNA testing to prove family links for those applying for an extended visa on the grounds of reuniting family members. The plan drew criticisms from civil rights groups and was modified so that DNA tests would be voluntary rather than mandatory and would be paid for by the French state. The first use of DNA profiling in a legal context related to an immigration case in the UK, where DNA testing proved that a young boy returning from Ghana was in fact a UK citizen.[506] DNA

504 European Commission Joint Research Centre, Institute for Prospective Technology Studies (2005) *op. cit.*

505 Law Reform Commission (2005) *op. cit.*

506 Newton G (2004) *DNA Fingerprinting Enters Society*. Available online at: http://genome.wellcome.ac.uk/doc_wtd020878.html, accessed 2 July 2009.

testing to satisfy immigration requirements is also used in the US and a number of EU countries.

Individuals are also making use of DNA as a personal identifier. DNA has been extensively used in determining paternity of children, while more innovative uses of DNA profiles are coming online. For paternity testing, samples are collected from both parents and the supposed offspring. Profiles are created and compared to check if the expected results are obtained, given that the genetic profile of the offspring should contain a contribution from both parents. Indeed, in the US, personal archival kits are sold that will allow conservation of a DNA sample. This enables an individual to be identified in the case of kidnapping, accidents or natural disaster, and allows families to identify their loved ones' remains.

Research involving single nucleotide polymorphisms (SNPs) is another area where DNA could potentially be used as an identifier. SNPs are predominantly used in research and to help understand inherited variation. An SNP occurs when a single nucleotide[507] in a genome sequence is altered and this variation is present in at least 1 per cent of the population.[508] Comparison of a number of SNPs could potentially be used as an identifier.

Critical Profile of DNA Recognition

DNA is present in all individuals and the structure of an individual's DNA does not change over time. In addition, DNA is highly distinctive because it is unique for everyone except identical twins.[509] DNA samples can be collected from all individuals, *i.e.* the FTE for this technology is zero. However, automatic, real time recognition is not, currently, possible because DNA matching requires chemical analysis, which takes a number of hours although it is envisaged that processing times will decrease in the future.[510,511] Apart from the inability to distinguish identical twins, the performance of DNA matching is highly accurate, though for certain applications the results need to be corroborated by human experts, which could leave it open to abuse or error. While the implementation of laboratory protocols and safeguards can help to reduce such problems, the technology is still sensitive to sample contamination and degradation, which can impinge on performance. The opportunity to deliberately contaminate a sample can be reduced through adequate supervision of the collection process.

A more serious issue for all technologies that utilise DNA (whether for forensics, paternity testing or research) relates to controlling the access to and storage of the DNA samples and DNA profiles. As noted above, privacy and security concerns surrounding the collection and storage of DNA could severely diminish the acceptability of the use of DNA for biometric

507 A nucleotide is a subunit of DNA consisting of a nitrogenous base (adenine, guanine, thymine or cytosine), a phosphate molecule and a sugar molecule (*i.e.* deoxyribose). Thousands of nucleotides are linked to form a DNA molecule.

508 For more information, see the Human Genome Project website: http://www.ornl.gov/sci/techresources/Human_Genome/home.shtml, accessed 2 July 2009.

509 It has been suggested that future advances in the technology could eventually enable the DNA of identical twins to be distinguished.

510 European Commission Joint Research Centre, Institute for Prospective Technology Studies (2005) *op. cit.*

511 Jain *et al.* (2004) *op. cit.*

recognition. While DNA profiling is based on the use of non-coding repetitive regions of DNA (which traditionally have been considered not to contain sensitive information about the individual), the DNA profiles themselves are generated from a full sample of the individual's DNA. This original DNA sample contains sensitive information about an individual, such as his/her susceptibility to particular diseases, and concerns have been raised regarding the potential misuse of this information.[512,513] The potential for function creep (the use of DNA samples for purposes beyond those envisaged at the time of collection) among DNA samples and/or data collected initially for law enforcement or research purposes is significant. Transfer of genetic information derived from DNA samples to third parties such as insurance companies or employers could, for example, lead to discriminating measures against individuals with a particular genetic make-up. It has been argued that both European and national regulations currently offer inadequate protection to completely prevent function creep.[514]

As previously mentioned, genetic information used in DNA identification (polymorphisms in the non-coding repetitive regions of DNA) was not thought to yield any information relating to specific traits or predispositions. Therefore, the information gleaned from DNA profiles could not be used in ways that would exceed that of individual identification. However, more recently it has become clear that information relating to race, ethnicity and familial relations can be derived from the non-coding regions of DNA. For example, the British Forensic Science Service has for some time been actively pursuing the possibility of predicting physical characteristics of individuals from their DNA profiles. They maintain that they have the capacity to discern, with unknown degrees of certainty, certain hair colour and ethnic origin.[515] Further research is being carried out in order to enable identification of height, eye colour and facial characteristics from DNA profiles. Finally, it should also be noted that information generated from DNA profiles does not simply relate to one individual. The closer the biological relationship between two people, the greater the chance of gaining information about person A by analysing the DNA profile of person B, *i.e.* DNA can be used to probabilistically identify family members, which raises ethical concerns when viewed in the larger sociological context.

Multimodal Biometrics

As noted above, unimodal biometric systems are susceptible to a number of different errors and limitations, for example, problems with noise in the captured biometric data, intra-class variation, limited distinctiveness of the modality, inter-class similarities, non-universality and the susceptibility to spoof attacks. Multimodal biometric systems represent the possibility of overcoming many of these problems by offering increased performance and accuracy, reliability, flexibility, inclusiveness and security.[516]

512 Nuffield Council on Bioethics (2007) *op. cit.*

513 It should be noted that in forensic applications the sensitive coding information from the DNA sample is discarded once the DNA profile and fingerprint have been produced.

514 Van Camp N and Dierickx K. (2008). The retention of forensic DNA samples: a socio-ethical evaluation of current practices in the EU. *Journal of Medical Ethics* 34(8): 606–610.

515 Lowe AL, Urquhart A, Foreman LA and Evett IW (2001). Inferring ethnic origin by means of an STR profile. *Forensic Science International* 119(1): 17–22.

516 Jain *et al.* (2004) *op. cit.*

What is Multimodal Biometrics?

In simplistic terms multimodal biometric systems use a number of biometric modalities from the same individual in the recognition process. Multimodal systems can be designed to work in five different ways (see Table 2).[517,518,519]

Table 2: Modes of Operation for Multimodal Systems

Design 1	Multiple sensors can be used to collect the same biometric.
Design 2	Multiple biometric modalities can be collected from the same individual, *e.g.* fingerprint and face, which requires different sensors.
Design 3	Multiple units of the same biometric are collected, *e.g.* fingerprints from two or more fingers.
Design 4	Multiple readings of the same biometric are collected during the enrolment and/or recognition phases, *e.g.* a number of fingerprint readings are taken from the same finger.
Design 5	Multiple algorithms for feature extraction and matching are used on the same biometric sample.

It is generally considered that multimodal systems incorporating a combination of independent modalities offer better performance than systems combining dependent modalities.[520] For this reason, Designs 2 and 3 should show the best accuracy, followed by Designs 4 and 5.

In addition, the design of multimodal systems also needs to consider at what stage the information from the modalities used should be combined, *i.e.* fused. For example, fusion could occur at the feature level, whereby features extracted from each of the modalities are used to produce a new, more distinctive template.[521] Alternatively, a matching score could be produced for each modality and these scores could be combined before the final decision is given. The last level of fusion can come at the decision stage, where a recognition decision is produced for each modality and a majority vote scheme makes the overall decision.[522,523] Fusion at the decision level could lead to problems if the recognition decisions from each of the modalities differ. Fusing the modality information earlier in the system is thought to result in better performance, as more biometric information is retained, upon which the recognition decision can then be based. However, this type of system is more difficult to design and most multimodal systems involve fusion at the matching score level.[524]

---

517 *ibid.*

518 Howells L (2005). *Fusion Comes in From the Cold.* A Consult Hyperion White Paper. Consult Hyperion, Surrey, 15p. Available online at: http://www.chyp.com/PubWebFiles/whitepaper/fusion_comes_in_from_cold.pdf, accessed 7 February 2008.

519 Multimodal systems can also use different combinations of the five design scenarios.

520 Jain *et al.* (2004) *op. cit.*

521 *ibid.*

522 European Commission Joint Research Centre, Institute for Prospective Technology Studies (2005) *op. cit.*

523 Howells (2005) *op. cit.*

524 Jain *et al.* (2004) *op. cit.*

Applications of Multimodal Biometrics

The ICAO standards for machine readable travel documents state that facial images must be used in all such documents, but these standards offer the scope to also include fingerprint and iris images.[525] For example, numerous countries within the EU and Europe,[526] which have signed up to the Schengen *acquis*, to facilitate easier travel between the assigned countries and into any country in the Schengen area from outside this zone, also require biometric passports.[527] It was originally decided under these regulations that all citizens of the participating countries must have biometric passports containing a facial image and two fingerprints (*i.e.* multimodal biometric passports) by June 2009. However, more recent regulations established in May 2009 have extended this deadline for the inclusion of fingerprints in biometric passports until June 2012.[528] It should be noted that, currently, Ireland and the UK are not part of the Schengen *acquis*.

In the US, the US Visitor and Immigrant Status Indicator Technology (US-VISIT) programme, which was introduced following the events of September 11th 2001, is multimodal. Under this system all visitors and immigrants to the US must scan all ten of their fingers and pose for a photograph, and this information is then checked against specific security databases and watch lists.[529] Enrolling ten fingerprints will help to increase the interoperability between US-VISIT and the FBI's (Federal Bureau of Investigation) fingerprint database (IAFIS).[530,531,532] The implementation of the US-VISIT programme was one of the principle reasons behind the switch to biometric passports in Ireland (and other European countries) to enable Irish citizens to continue to travel to the US as part of a visa waiver scheme. In addition, the US Department of Homeland Security (DHS) intends to implement procedures for the collection of biometric information from foreign nationals as they exit the US as part of the US-VISIT programme.[533]

In Ireland, the proposed Immigration, Residence and Protection Bill 2008,[534] if implemented, could require all legal migrants hoping to enter Ireland to provide

525 ICAO TAG (2004) *op. cit.*

526 The countries currently involved are Austria, Belgium, Czech Republic, Denmark, Estonia, Finland, France, Germany, Greece, Hungary, Italy, Latvia, Lithuania, Luxembourg, Malta, Netherlands, Poland, Portugal, Slovakia, Slovenia, Spain and Sweden, as well as Iceland, Norway and Switzerland.

527 Council Regulation (EC) No 2252/2004 of 13 December 2004 on standards for security features and biometrics in passports and travel documents issued by Member States.

528 Regulation (EC) No 444/2009 of the European Parliament and of the Council of 28 May 2009 amending Council Regulation (EC) No 2252/2004 on standards for security features and biometrics in passports and travel documents issued by Member States.

529 For more information see the Department of Homeland Security website, available online at: http://www.dhs.gov/files/programs/editorial_0525.shtm, accessed 21 July 2009.

530 Department of Homeland Security (2007). *Privacy Impact Assessment Update for the Conversion to 10-Fingerprint Collection for the United States Visitor and Immigration Status Indicator Technology Program (US-VISIT)*. Department of Homeland Security, Washington, 15p. Available online at: http://www.dhs.gov/xlibrary/assets/privacy/privacy_pia_usvisit_10p.pdf, accessed 22 May 2008.

531 Cavoukian and Stoianov (2007) *op. cit.*

532 However, once all ten fingerprints have been enrolled, it is considered unlikely that all ten fingerprints will need to be collected, at each subsequent encounter, in order to verify an individual's identity.

533 For more information see the Department of Homeland Security website, available online at: http://www.dhs.gov/files/programs/editorial_0525.shtm, accessed 21 July 2009.

534 Immigration, Residence and Protection Bill (2008).

a photograph, all ten fingerprints and both palm prints as part of their application procedure, although the exact biometric information to be included has not been finalised. This information would then be included on that individual's residency permit. This system would also be connected to other European databases for cross-referencing purposes.

The UK have already implemented a multimodal biometric immigration system, which involves taking a photograph and ten fingerprints from all immigration applicants.[535,536] These are then checked against the criminal database in the UK for any potential matches. It should be noted that anyone unwilling to provide their face and fingerprint biometrics cannot apply for a visa to the UK.

Critical Profile of Multimodal Systems

On the whole, the use of independent modalities in multimodal systems is more reliable and improves system accuracy.[537,538] In addition, combining the modalities in the appropriate sequence can improve the system's overall accuracy and performance. For example, in a large-scale identification application, facial recognition could be used first to quickly produce a series of possible matches, fingerprint recognition could then be used to single out the correct individual from this series.[539] Multimodal systems can counteract the non-universality of one biometric modality by combining it with another, which facilitates greater coverage and inclusion of the potential user population.[540] This provides increased flexibility to the system, which helps to lower the failure to enrol rate (FTE).[541] Multimodal systems are reputed to offer increased security because they require greater effort to defeat the system.[542,543] For example, an individual could be required to provide a random subset of the enrolled biometric modalities (*i.e.* challenge-response authentication), which is more difficult to spoof.

Nonetheless, further research and independent testing of multimodal system performance and design are required, in order to extrapolate current results to large-scale applications.[544] Moreover, decisions on whether or not to implement a specific multimodal system need to consider the level of performance and accuracy required, degree of usability and flexibility it

535 For more information see the UK Border Agency website, available online at: http://www.ukvisas.gov.uk/en/howtoapply/biometricvisa/, accessed 22 July 2009.

536 Some individuals are exempt from the requirement to provide their biometric information, for example Heads of State and children under six years of age.

537 Jain *et al.* (2004) *op. cit.*

538 European Commission Joint Research Centre, Institute for Prospective Technology Studies (2005) *op. cit.*

539 Hong L and Jain AK (1999). Multimodal Biometrics. In A Jain, R Bolle and S Pankanti (eds.) *Biometrics: Personal Identification in Networked Society*, Kluwer Press, Dordrecht, p.327–344.

540 European Commission Joint Research Centre, Institute for Prospective Technology Studies (2005) *op. cit.*

541 Howells (2005) *op. cit.*

542 Bowyer KW, Chang KI, Yan P, Flynn PJ, Hansley E and Sarkar S (2006). *Multi-Modal Biometrics: An Overview*. Second Workshop on Multi-Modal User Authentication (MMUA 2006), May 2006, Toulouse, France, 8p. Available online at: http://www.nd.edu/~kwb/BowyerEtAlMMUA_2006.pdf, accessed 22 February 2008.

543 Howells (2005) *op. cit.*

544 Bowyer *et al.* (2006) *op. cit.*

can offer, and the computational and logistical resources needed, as well as the overall cost, particularly when compared to unimodal systems.

There appears to be substantial scope for combining existing and emerging unimodal biometric technologies in multimodal systems, for example, 2D and 3D face, voice, face and lip movement, and fingerprint and face.[545]

Future Biometric Modalities

Notwithstanding future developments and advancements with established and multimodal biometric technologies, a number of other modalities are currently being investigated as potential biometric identifiers. For example, research has indicated that an individual's baseline brainwave pattern from electroencephalogram (EEG) recordings is distinctive enough to be used as a means of biometric recognition.[546] While EEG patterns have been assessed for use in identification-based systems, this modality appears to perform better in the verification mode.[547] Electrocardiograms (ECGs) are also considered to have potential from the point of view of biometric recognition, both in unimodal and multimodal biometric systems.[548] Footprint recognition, focusing on foot geometry, shape and texture,[549] as well as foot pressure and distribution is also being investigated for the purposes of identification and verification.[550] In addition, the geometric shape and physiological structure of the tongue has been analysed particularly for verification-based applications.[551] While these and other potential modalities have shown promise during preliminary research, they still need to be assessed in greater detail against the seven pillars of biometrics (universality, distinctiveness, permanence, collectability, performance, acceptability, and resistance to circumvention), particularly if they are intended to be used in large-scale applications.

545 Additional examples of potential multimodal combinations include 2D face and ear geometry; fingerprint and hand geometry; fingerprint and palm print; face and voice; *etc.*

546 Riera A, Soria-Frisch A, Caparrini M, Grau C and Ruffini G (2008). Unobtrusive Biometric System Based on Electroencephalogram Analysis. *EURASIP Journal on Advances in Signal* Processing Volume 2008: Article ID 143728, 8p, doi:10.1155/2008/143728.

547 *ibid.*

548 Chan ADC, Hamdy MM, Badre A and Badee V (2006). Person Identification Using Electrocardiograms. *Canadian Conference on Electrical and Computer Engineering 2006 (CCECE '06)* May 2006: 1–4.

549 Uhl A and Wild P (2008). Footprint-based biometric verification. *Journal of Electronic Imaging* 17(1): 11–16.

550 Heyer R (2008) *op. cit.*

551 Zhang D, Liu Z, Yan J and Shi P (2007). Tongue-Print: A Novel Biometrics Pattern. *Lecture Notes in Computer Science* 4642: 1174–1183.

CHAPTER 3
ETHICAL CONSIDERATIONS FOR BIOMETRIC INFORMATION AND ITS ASSOCIATED TECHNOLOGIES

Chapter 3: Ethical Considerations for Biometric Information and its Associated Technologies

Considering the developments in biometric technologies, the increasing incidences of their deployment and the diversity of their applications, it is imperative that the ethical, social and legal issues surrounding the use of biometrics are examined and discussed.[552,553] Similarly to other developments in science and technology, the challenges posed are not with the use of biometric technologies *per se*, but in how they are applied and how the resulting data is dealt with. Furthermore, when considering the ethical issues associated with biometric applications and systems it is important not to consider these developments in isolation, but rather in conjunction with other innovations, such as increased information accessibility and surveillance as well as network and database connectivity.[554]

The use of biometric systems and applications raises a number of ethical questions, particularly relating to basic rights such as privacy, autonomy and bodily integrity. While these rights are legally protected both nationally and internationally,[555] concerns have been raised that these rights could be disproportionately superseded under the guise of acting for the common good, for example, for national security or public health and safety.[556,557,558,559] While there certainly are cases where an individual's rights may have to be sacrificed for the common good, this should not be the default position. Accordingly, the introduction of any biometric technology needs to be both proportional and transparent.

The majority of ethical concerns raised with regard to the development and implementation of biometric technologies relate to privacy.[560,561,562] Indeed, the perceived intrusion the use of biometrics could have on individual privacy is often considered as one of the main barriers to the wider acceptance of these technologies.[563,564,565]

---

552 National Consultative Ethics Committee for Health and Life Sciences, 2007. *Biometrics, identifying data and human rights.* Opinion No. 98. National Consultative Ethics Committee for Health and Life Sciences, France, 22p.

553 For more information see the Biometric Identification Technology Ethics (BITE) Project website, available online at: http://www.biteproject.org/, accessed 16 October 2007.

554 van der Ploeg I (2005b). *The Politics of Biometric Identification. Normative aspects of automated social categorization.* BITE Policy Paper No.2. 16p. Available online at: http://www.biteproject.org/documents/politics_of_biometric_identity%20.pdf, accessed 8 November 2007.

555 For example, the rights listed above a variously protected under the Constitution of Ireland (1937), the Convention on Human Rights and Biomedicine (1997), the Charter of Fundamental Rights of the European Union (2000) and the UN Universal Declaration of Human Rights (1948).

556 National Consultative Ethics Committee for Health and Life Sciences (2007) *op. cit.*

557 International Commission of Jurists (2009). *Assessing Damage, Urging Action. Report of the Eminent Jurists Panel on Terrorism, Counter-terrorism and Human Rights.* International Commission of Jurists, Geneva, 199p.

558 Alterman A (2003). "A piece of yourself": Ethical issues in biometric identification. *Ethics and Information Technology* 5(3): 139–150.

559 Clarke DM (1984). *Church and State: Essays in Political Philosophy.* Cork University Press, Cork, 275p.

560 Snijder (2007) *op. cit.*

561 NSTC Subcommittee on Biometrics (2006d) *op. cit.*

562 Woodward (2001) *op. cit.*

563 Ponemon L (2006). *Global Study on the Public's Perception about Identity Management.* Ponemon Institute and Unisys Corporation, Michigan, US, 27p.

564 European Commission Joint Research Centre, Institute for Prospective Technology Studies (2005) *op. cit.*

565 Woodward (2001) *op. cit.*

In advance of a recent general election in Togo, fingerprint and face recognition-based biometric authentication were used to create a secure database of registered voters and to produce unique voter cards. The system was implemented to reduce problems with multiple voters.[566] Fingerprint recognition has also been used elsewhere in Africa (Malawi and Mozambique) by the Opportunity International Bank of Malawi to enable those individuals who may not be literate or who lack formal identification documents to avail of banking and financial services that would otherwise not have been accessible to them.[567,568] A similar initiative, also involving fingerprint recognition, was established in Mexico by Banco Azteca and provides banking services to individuals who would otherwise not have been given access to such facilities because they lacked reliable means of identification.[569,570]

Privacy

Why is Privacy Important?

Privacy is often viewed as a fundamental right,[571,572,573] so much so that it is considered one of the most important human rights of the modern age.[574] The importance of privacy is underlined by the fact that it is recognised and respected in different cultures throughout the world and it is protected in a multitude of national and international treaties, conventions and constitutions. However, despite the widely held respect for, and protection of, the right to privacy, there is still some difficulty in defining privacy as a concept.[575,576]

Privacy holds an intrinsic importance for many people, *i.e.* the idea that privacy should be valued for its own sake, even though they may not be able to quantify exactly why privacy is important to them. Many people retain a "sense of privacy", *i.e.* an understanding that certain aspects of their lives are no one else's business, but their own.[577] This view is perpetuated

---

566 Rommelaere J (2007). *Togo government accomplishes nationwide biometric registration of its voters.* Presentation at the Biometrics Exhibition and Conference 2007, 17-19 October 2007, Westminster, London.

567 MacDonald F (2008). A card up Africa's sleeve. *Metro* 26 February 2008.

568 For more information see the Opportunity International Canada website, available online at: http://www.opportunityinternational.ca/learn/current.html, accessed 27 July 2009.

569 Biometric Technology Today (2006). Banco Azteca rolls out biometrics to 8m customers. *Biometric Technology Today* 14(5): 4.

570 Digital Persona, Inc. (2006). *Digital Persona Deploys World's Largest Biometric Banking Application.* Available online at: http://www.digitalpersona.com/index.php?id=pr_20060322, accessed 12 January 2009.

571 Electronic Privacy Information Center and Privacy International (2007). *Privacy and Human Rights 2006. An International Survey of Privacy Laws and Developments.* Electronic Privacy Information Center and Privacy International, US. Available online at: http://www.privacyinternational.org/article.shtml?cmd[347]=x-347-559458, accessed 22 September 2008.

572 OECD, Working Party on Information Security and Privacy (2004) *op. cit.*

573 Reiman JH (1984). Privacy, Intimacy, and Personhood. In FD Schoeman (ed.) *Philosophical Dimensions of Privacy. An Anthology.* Cambridge University Press, New York, p.300–316.

574 Electronic Privacy Information Center and Privacy International (2007) *op. cit.*

575 *ibid.*

576 Thomson JJ (1984). The Right to Privacy. In FD Schoeman (ed.) *Philosophical Dimensions of Privacy. An Anthology.* Cambridge University Press, New York, p.272–289.

577 Rachels J (1984). Why Privacy is Important. In FD Schoeman (ed.) *Philosophical Dimensions of Privacy. An Anthology.* Cambridge University Press, New York, p.290–299.

through the frequent descriptions of the concept of privacy as an individual's right to be left alone or a barrier against intrusion from the outside world.[578,579,580,581]

Numerous attempts have been made to elucidate the basis of privacy and, thus, determine why it is so important to us. In his analysis of privacy, Charles Fried highlighted the importance of privacy as a right with the suggestion that invasions of people's privacy "injure them in their very humanity".[582] In her discussion of privacy, Judith Jarvis Thomson proposes that the right to privacy is derived from other rights, particularly an individual's right over his/her person (body) and his/her property.[583] However, Thomson's concept of privacy as a derivative of other rights has been criticised. Jeffrey H. Reiman contends that an individual's right over his/her person and his/her property are expressions of the right to privacy and are, in fact, derived from it, not the other way round.[584] Moreover, Reiman suggests that Thomson's theory of privacy as a minor aspect of other personal and property rights downplays the actual value of privacy. He goes on to suggest that the right to privacy protects some unique interest of ours that goes beyond the degree of protection offered by personal and property rights.[585]

James Rachels also criticises Thomson's privacy hypothesis as inadequate because he suggests that situations could arise where an individual's right to privacy could be violated without violating either his/her rights over his/her person or his/her property.[586] Rachels gives the example of someone finding out very personal information about an individual (*e.g.* that he is impotent) and passing this information on to other people. He argues that such a scenario would not violate that individual's rights over his/her person or his/her property rights, but it would still be a violation of his/her right to privacy.[587] Therefore, while an individual's right over his/her person or property are important rights, which can be connected to privacy, these rights do not always overlap with the right to privacy. Similarly to Reiman, Rachels suggests that an individual's right to privacy should be valued in its own right because it protects some other special interest.[588]

Therefore, the question arises: what is this special interest that makes the right to privacy so important to us as individuals? Rachels proposes that the value of privacy is derived from the notion that "there is a close connection between our ability to control who has access to us and to information about us, and our ability to create and maintain different sorts of social relationships with different people".[589]

---

578 Moor JH (1990). The Ethics of Privacy Protection. *Library Trends* 39(1 and 2): 69–82.

579 OECD, Working Party on Information Security and Privacy (2004) *op. cit.*

580 Rachels J (1984) *op. cit.*

581 Electronic Privacy Information Center and Privacy International (2007) *op. cit.*

582 Fried C (1984). Privacy [A moral analysis]. In FD Schoeman (ed.) *Philosophical Dimensions of Privacy. An Anthology.* Cambridge University Press, New York, p.203–222.

583 Thomson (1984) *op. cit.*

584 Reiman (1984) *op. cit.*

585 *ibid.*

586 Rachels (1984) *op. cit.*

587 *ibid.*

588 *ibid.*

589 *ibid.*

Globalisation with its associated digital and networked society has greatly increased the opportunity, scope and ease of identity theft.[590] Identity theft has been classified as the fastest growing white collar crime since the mid-1990s.[591] It has been estimated that approximately 20 per cent of the US population have been the victim of identity theft.[592,593] In 2007 alone about 8.4 million people in the US were victims of identity theft, which cost over $49 billion.[594] Moreover, in the UK it is estimated that over 100,000 people are affected by identity theft every year,[595] at a cost of £1.2 billion.[596] The Data Protection Commissioner of Ireland has suggested that the lack of a unique identifier, similar to the US social security number, may be partially responsible for the lower incidences of identity theft here.[597] Nonetheless, a survey conducted in October 2008 revealed that 87,000 people in Ireland have been the victim of identity theft.[598] As an example of this one person was defrauded of €22,000 by an imposter using a forged passport.[599] Biometric technologies are suggested to offer increased protection against the problems of identity theft. In fact, the potential of biometric modalities to combat the problem of identity theft is one of the main drivers behind the implementation of many biometric applications.

From a privacy perspective biometric technologies can impact positively or negatively on an individual. Biometric technologies can provide an accurate and rapid method of identification, thereby enhancing privacy and security – for example, by helping to secure personal information, by assisting an individual to retain control over his/her own information and by reducing the likelihood of identity theft.[600] However, the use of biometric technologies may also threaten an individual's privacy, and this technology has been criticised for its perceived Orwellian, invasive potential.[601,602,603] The use of biometric information raises concerns about

590 For the purposes of this document the term identity theft will be used for both identity theft and identity fraud.

591 Cavoukian A (2005). *Identity Theft Revisited: Security is Not Enough.* Information and Privacy Commissioner/ Ontario, Toronto, 39p. Available online at: http://www.ipc.on.ca/images/Resources/idtheft-revisit.pdf, accessed 7 February 2008.

592 *ibid.*

593 The Chubb Corporation (2005). *One in Five Americans Has Been a Victim of Identity Fraud.* Available online at: http://www.chubb.com/corporate/chubb3875.html, accessed 12 January 2009.

594 Privacy Rights Clearinghouse (2007). *How Many Identity Theft Victims Are There? What Is the Impact on Victims?* Available online at: http://www.privacyrights.org/ar/idtheftsurveys.htm#FTC, accessed 16 June 2009.

595 Davies S, Hosein I and Whitley EA (2005). *The Identity Project: An assessment of the UK Identity Cards Bill and its implications.* The London School of Economics and Political Science London, 303p. Available online at: http://eprints.lse.ac.uk/684/1/identityreport.pdf, accessed 15 February 2008.

596 For more information see the Home Office Identity Fraud Steering Committee website: http://www.identitytheft.org.uk/faqs.asp, accessed 16 June 2009.

597 Data Protection Commissioner (2009). *Twentieth Annual Report of the Data Protection Commissioner 2008.* 111p. Available online at: http://www.dataprotection.ie/documents/annualreports/AR2008.pdf, accessed 14 May 2008.

598 Michael J (2008). Identity theft crime affects 87,000 – survey. *The Irish Times* published online 6 October 2008. Available online at: http://www.irishtimes.com/newspaper/breaking/2008/1006/breaking23.htm, accessed 15 October 2008

599 RTÉ Business (2005). *Bank customers warned to be on their guard.* RTÉ Business published online 9 November 2005. Available online at: http://www.rte.ie/business/2005/1109/ipso.html, accessed 17 June 2009.

600 Biometric Information Technology Ethics (2005). *Biometrics and Privacy.* Report of the Second BITE Scientific Meeting, Tuesday 26th April 2005, Rome, Italy, 13p. Available online at: http://www.biteproject.org/documents/report_biometrics_privacy.pdf, accessed 16 October 2007.

601 Woodward *et al.* (2001) *op. cit.*

602 Lodge J (2007). Freedom, security and justice: the thin end of the wedge for biometrics? *Annali dell Institute Superiore di Sanitá* 43(1): 20–26.

603 Etzioni A (1999). *The Limits of Privacy.* Basic Books, New York, 288p.

the ability of an individual to control the information about him/herself that he/she is willing to make available to others, which would necessarily impact on his/her right to privacy. The privacy concerns related to biometrics are manifest in two spheres, those relating to personal privacy (*i.e.* fears about the erosion of personal identity and bodily integrity) and those relating to informational privacy (*i.e.* fears about the misuse of data and function creep).

Personal Privacy

Our biometric information has the ability to"uniquely" identify us. Indeed, this specific feature of biometric information is one of the reasons that these technologies tend to evoke such heightened privacy concerns. Anton Alterman argues that, because biometric images facilitate our identification, we have a fundamental interest in controlling their creation and use.[604] Alterman believes that morally we have a greater interest in body-based information owing to the relationship between our body and our conception of self.[605] Different experiences and interactions feed into this sense of self, which engenders a degree of complexity to each individual's personal identity. The ability to maintain and develop this complex identity is facilitated by, and is, thus, interconnected with, our possession of personal rights, such as autonomy, bodily integrity, and, particularly, privacy.

Informatisation of the Body

Biometrics are considered, by many people, to provide the optimum form of identification because as physical, physiological or behavioural characteristics they represent "something you are". Biometrics are, therefore, considered to offer stronger assurances of identifying individuals correctly than was previously possible through more traditional identification methods,[606,607] (*e.g.* passwords, PINs, birth certificates, passports, national identity cards and other photographic identity cards). Governments, companies and organisations now require more stringent confirmation of people's identities for numerous reasons, for example, for national border control, to combat crime, fraud and terrorism, to control access to services, and to control physical and logical access. Therefore, given the increasing necessity for secure identity authentication, biometric recognition could become the default form of identity, ultimately being required in all instances where an individual may need to be recognised.[608,609]

604 Alterman (2003) *op. cit.*
605 *ibid.*
606 European Commission Joint Research Centre, Institute for Prospective Technology Studies (2005) *op. cit.*
607 OECD, Working Party on Information Security and Privacy (2004) *op. cit.*
608 Woodward *et al.* (2001) *op. cit.*
609 European Commission Joint Research Centre, Institute for Prospective Technology Studies (2005) *op. cit.*

Numerous criteria have been used throughout history as the basis of classifying an individual's identity, for example, gender, ethnicity, race, religion, class, nationality, *etc.*[610] Traditionally, an individual's identity was distinguished through the use of attributed and biographical characteristics,[611,612] for example, his/her name and occupation. However, the processes of industrialisation and urbanisation have led to the development of large-scale societies, increased mobility of the population and, ultimately, the advent of nation-states. These large-scale societies brought with them an increasing need to clarify exactly who each individual was. Consequently, identity started to be confirmed on the basis of an individual's name and, subsequently through the use of identifying documents such as birth certificates, passports and national identity cards.[613,614] Societies have continued to progress beyond the idea of the large-scale national society to an international and globalised community, interconnected through advances in transportation, communications and information technologies. These networked technologies are challenging conventional ideas about identity and identification, which has resulted in a greater need for individuals to prove their identity.[615] As a result, traditional forms of identification and recognition are no longer considered wholly adequate.[616] This is particularly relevant in developing countries, where many people have weak and unreliable identity documents or none at all.[617] As part of the search for more robust and "unique" forms of identification, biometric features and traits are increasingly being used as a means of recognising individuals.

With the predicted increase in the use of biometric technologies in the future, there are concerns that people will be recognised more and more solely on the basis of their biometric information.[618,619] Identity could, therefore, become grounded in the physical body, *i.e.* the body, in the form of the biometric, would become the password.[620,621,622] In this regard, biometric-based identification has been likened to a compulsory identity card that is effectively

610 Mordini E (2008). Biometrics, Human Body, and Medicine: A Controversial History. In P. Duquenoy, C. George and K. Kimppa (eds.) *Ethical, Legal, and Social Issues in Medical Informatics.* IGI Global, London, p.249–272.

611 *ibid.*

612 For example, in small-scale societies, such as tribes and villages, an individual's identity was confirmed on the basis of physical and cultural appearance and location [Mordini E and Ottolini C (2007) *op. cit.*].

613 Mordini (2008) *op. cit.*

614 Albrecht A, Behrens M, Mansfield T, McMeehan W, Rejman-Greene M, Savastano M, Statham P, *et al.* (2003). *BIOVISION: Roadmap for Biometrics in Europe 2010.* Final Report of the Roadmap Task, D2.6/Issue 1.1, 205p. Available online at: http://ftp.cwi.nl/CWIreports/PNA/PNA-E0303.pdf, accessed 9 July 2008.

615 Mordini and Ottolini (2007) *op. cit.*

616 Mordini (2008) *op. cit.*

617 For example, it was calculated that, in 2000 41% of births worldwide (*i.e.* 50 million infants) were not registered and, therefore, had no identifying documentation. Moreover, in a number of countries, including Pakistan, Bangladesh and Nepal child registration is not yet compulsory (Mordini [2008] *op. cit.*).

618 Albrecht *et al.* (2003) *op. cit.*

619 van der Ploeg I (2005a). *Biometric Identification Technologies: Ethical Implications of the Informatization of the Body.* BITE Policy Paper No.1. 18p. Available online at: http://www.biteproject.org/documents/policy_paper_1_july_version.pdf, accessed 16 October 2007.

620 Biometric Information Technology Ethics (2005) *op. cit.*

621 Mordini E and Massari S (2008) *op. cit.*

622 van der Ploeg I (1999). The illegal body: 'Eurodac' and the politics of biometric identification. *Ethics and Information Technology* 1(4): 295–302.

glued to the body or that the body itself becomes the identity card.[623] Concerns have been expressed about this redefinition of the body in terms of identifying information, that is, the informatisation of the body.[624,625,626] It has been argued that individuals are becoming characterised by the aggregated pieces of biometric (or other personal) information relating to them, rather than as individuals in their own right.[627,628,629]

Moreover, characterising and processing people as information could be considered equivalent to treating people as mere objects, enabling them to be used as a means to another's ends, for example, for processing and categorisation.[630,631] Irma van der Ploeg has argued that the individual, through his/her body, becomes a machine readable item and his/her identity is determined on the basis of this reading.[632] Kant considered the objectification of individuals a violation of human dignity.[633] Others have since deemed the informatisation of the body to be incompatible with the principle of human dignity.[634,635]

The body has, in the past, been used as a means of political control, with people in particular groups or categories being "branded" or labelled for identification purposes, for example, criminals in late ancien régime in France, or Nazi prisoners during World War II.[636] Such body-based identification was considered dehumanising by many people.[637,638] However, similar concerns have also arisen about more recent methods of control involving biometric technologies, for example, the philosopher Giorgio Agamben has criticised the collection of biometric information as part of the US-VISIT immigration programme.[639] He suggests that while such mechanisms of control would previously have been considered inhumane and extraordinary, they are now being proffered as normal and routine and could potentially "be the precursor to what we will be asked to accept later as the normal identity registration of a good citizen in the [s]tate's gears and mechanisms".[640] For this reason he believes such methods of control should be opposed now.[641]

---

623 *ibid.*

624 National Consultative Ethics Committee for Health and Life Sciences (2007) *op. cit.*

625 van der Ploeg (2005a) *op. cit.*

626 Biometric Information Technology Ethics (2005) *op. cit.*

627 Albrecht *et al.* (2003) *op. cit.*

628 National Consultative Ethics Committee for Health and Life Sciences (2007) *op. cit.*

629 van der Ploeg (2005a) *op. cit.*

630 Alterman (2003) *op. cit.*

631 National Consultative Ethics Committee for Health and Life Sciences (2007) *op. cit.*

632 van der Ploeg (2005a) *op. cit.*

633 Immanuel Kant, the eighteenth-century philosopher, defined a concept of human dignity, which demands equal respect for all persons based on their capacity for rational autonomy. Kant's conception prohibited the use of persons merely as a means to another person's ends.

634 Alterman (2003) *op. cit.*

635 National Consultative Ethics Committee for Health and Life Sciences (2007) *op. cit.*

636 Mordini (2008) *op. cit.*

637 *ibid.*

638 Mordini and Ottolini (2007) *op. cit.*

639 Agamben G (2004). No to Bio-Political Tattooing. *Le Monde* 10 January 2004.

640 *ibid.*

641 *ibid.*

Social categorisation, particularly where it involves the use of biometric information, could also be seen as a form of "branding" in that a specific identity is ascribed to the person for the purposes of placing them in a particular category, for instance, immigrant, suspect, potential criminal, or potential terrorist, which can then be used for the purposes of social control. For example, immigration procedures at an airport may use biometrics to authenticate an individual's identity and compare them with various databases. As a result, that individual could be categorised as "known" or "unknown", "legal" or "illegal", "wanted" or "unwanted", or a "low" or "high security risk".[642] Once assigned to a specific category, it can prove difficult for an individual to rid him/herself of that ascribed identity, even where it is shown to be inaccurate.[643,644] This issue is particularly pertinent when the ascribed identity is linked to some biometric characteristic. Irma van der Ploeg has suggested that such ascribed identities would effectively become like an individual's shadow: "hard to fight, impossible to shake".[645] Consequently, ascribing an identity and classifying people in such ways impacts not only on the individual him/herself, but also on the way in which society reacts to that person,[646,647] particularly in cases where such an identity is ascribed erroneously. While such misidentifications are considered to be less likely, though certainly not impossible, with the use of biometrics, the possible over reliance on the veracity of biometric identification could make such mistakes harder to rectify, with potentially far reaching repercussions for the individuals involved.[648,649,650]

In the US in May 2004 Mr Brandon Mayfield was wrongfully arrested in connection with the terrorist attacks in Madrid in March of that year.[651,652] During their investigations the Spanish authorities found a fingerprint on evidence linked to the bombings. This fingerprint was examined by the FBI, who stated that it belonged to Mr Mayfield and he was arrested.[653,654] Mr Mayfield was only released two weeks later after the Spanish authorities had identified that the fingerprints actually belonged to another man, Mr Ouhnane Daoud.[655] A review of Mr Mayfield's case revealed that, while there were some similarities between his fingerprints and those found on the evidence, a number of errors and biases in the FBI examination contributed to the misidentification of Mr Mayfield.[656] The potential problem of over relying on biometric-based identification

642 van der Ploeg (2005b) *op. cit.*

643 *ibid.*

644 Article 29 Data Protection Working Party (2003) *op. cit.*

645 van der Ploeg (2005b) *op. cit.*

646 Henschke A (2007). *An Evaluation of Forensic DNA Databases Using Different Conceptions of Identity.* MSc Thesis, Linköping University, Sweden, 71p. Available online at: http://liu.diva-portal.org/smash/record.jsf?pid=diva2:23785, accessed 23 July 2008.

647 Albrecht *et al.* (2003) *op. cit.*

648 Graham-Rowe D (2005). Privacy and prejudice: whose ID is it anyway? *New Scientist* 187(2517): 20–23.

649 Lyon D (2008). Biometrics, Identification and Surveillance. *Bioethics* 22(9): 499–508.

650 van der Ploeg (2005b) *op. cit.*

651 New Scientist (2005). The myth of fingerprints. *New Scientist* 187(2517): 3.

652 Office of the Inspector General (2006). *A Review of the FBI's Handling of the Brandon Mayfield Case.* US Department of Justice, Washington, 330p. Available online at: http://www.usdoj.gov/oig/special/s0601/PDF_list.htm, accessed 12 January 2009.

653 Harden B (2004). FBI Faulted in Arrest of Ore. Lawyer. *The Washington Post* 16 November 2004. Available online at: http://www.washingtonpost.com/wp-dyn/articles/A52907-2004Nov15.html, accessed 12 January 2009.

654 Office of the Inspector General (2006) *op. cit.*

655 *ibid.*

656 *ibid.*

was also manifest in the case of Mr Rene Ramon Sanchez. Mr Sanchez was stopped in 1995 for a driving offence and had his fingerprints taken by police. However, while being processed, Mr Sanchez's fingerprints were incorrectly combined and stored with the personal details of Mr Leo Rosario (who was a known drug dealer and candidate for deportation).[657,658] This mistake resulted in Mr Sanchez being arrested and detained a number of times on arrest warrants issued for Mr Rosario. In each case, Mr Sanchez's efforts to clear his name were hampered due to the emphasis given to the fact that his fingerprints matched those on file for Mr Rosario (*i.e.* Mr Sanchez's own fingerprints).[659,660] On each occasion, Mr Sanchez was eventually released when other information indicated that he was not Mr Rosario (*e.g.* after comparing photographs of the two men). The error with Mr Sanchez's fingerprint records was finally resolved in 2002.[661]

The misidentification and "labelling" of individuals can be manifest in another way when dealing with biometric systems. As noted previously, biometric recognition is not totally accurate and errors do occur, including false reject errors.[662] In some cases such errors may be temporary, caused by intra-class variation[663] and the matching constraints of the system involved.[664] Such errors may cause some minor inconvenience and embarrassment to the individual involved, who will have to re-attempt the recognition process. However, fallback procedures, for instance, human supervision of the acquisition process, can be implemented in such cases to avoid an individual being excluded from a system he/she is rightly entitled to use or access. However, a more serious issue arises in relation to individuals who consistently experience problems in using biometric systems, for example, due to an injury, a medical condition (*e.g.* arthritis, cataracts), a physical, mental or learning disability, their age, or even their race.[665,666,667] David Lyon has stated that the failure to enrol rates (FTE) among non-white groups such as Hispanics, blacks and Asians are generally higher than among white people.[668] Joseph Pugliese argues that biometric technologies, specifically face-, fingerprint- and iris-based systems, are inherently biased towards white users, *i.e.* these systems are "infrastructurally calibrated to whiteness", which leads to the higher FTE rates among non-

---

657 *Sanchez* v. *The State of New York*, #2002-001-034, Motion No. M-64552.

658 Lyon (2008) *op. cit.*

659 Weiser B (2004). Can Prints Lie? Yes, Man Finds to His Dismay. *The New York Times* 31 May 2004. Available online at: http://www.nytimes.com/2004/05/31/nyregion/31IDEN.html, accessed 23 February 2009.

660 *Sanchez* v. *The State of New York*, *op. cit.*

661 *ibid.*

662 As noted above, a false reject error occurs when an acquired template from one individual does not match the enrolled template for that individual.

663 Intra-class variation relates to the fact that no two samples of the same biometric from the same person are ever absolutely identical and it can be caused by differences in a number of factors between both sample collection times, *e.g.* differences in the ambient conditions, imperfect imaging conditions, changes in the user's biometric characteristic or in the user's interaction with the sensor.

664 For example, the threshold may require a high degree of similarity between the enrolled biometric template and the comparison template.

665 For example, in the case of fingerprint recognition, elderly individuals may have thinner skin, which can result in poor image resolution.

666 Wickins J (2007). The ethics of biometrics: the risk of social exclusion from the widespread use of electronic identification. *Science and Engineering Ethics* 13(1): 45–54.

667 van der Ploeg (2005b) *op. cit.*

668 Lyon (2008) *op. cit.*

whites.[669] Lyon acknowledges that if such bias towards white users did indeed predominate in biometric systems, "it would clearly be a cause of serious concern".[670] Given the likelihood of problems with enrolment and acquistion, a case has been made regarding the need to establish systems that can handle these exceptions, for example, through the use of multiple and/or alternative biometric modalities, without disenfranchising the individuals involved and without creating a stigma on such individuals because they cannot use the "normal" biometric system.[671,672] The Institute for Prospective Technological Studies of the European Commission Joint Research Centre has expressed the view that those individuals who cannot use the system have the same need for dignity and security as those who can,[673] otherwise, these people could potentially become second-class citizens, who are discriminated against because their bodies do not conform to some preset biometric criterion.

Identity, Privacy and Bodily Integrity

The Council recognises that, as a social construct, the concept of a single identity has important public and legal ramifications for an individual's rights, responsibilities and accountability, for example, in relation to citizenship, judicial processes, healthcare situations, social welfare entitlements, owning private property, taxation systems and travel.[674] However, an individual's identity is a complex and multifaceted condition, which is both affected by, but also influences, that individual's experiences and relationships.[675] As such, identity is not static – it develops and evolves over time. It is said to be "processual, fluid and constantly in flux dependent on the social, political, economic and ideological aspects of the situations individuals … find themselves in".[676] Our interactions, experiences and relationships with other people help to orient us and contribute to our sense of self within society and our sense of belonging amongst others.[677,678] Accordingly, while an individual remains the same person (and body) his/her personal identity evolves and develops throughout his/her life. The French National Consultative Ethics Committee on Health and Life Sciences has argued that fixing a single identity to an individual based on particular physical or bodily characteristics could degrade this concept of personal identity. In this regard, identity could be said to be changing from an individual right to an obligation or social duty.[679] However, the notion of an integrated "single" identity is not, necessarily, accepted by everyone, and some people prefer to present different personas or different aspects of their identity, *i.e.* partial identities, in different social

---

669 Pugliese J (2005). *In Silico* Race and the Heteronomy of Biometric Proxies: Biometrics in the Context of Civilian Life, Border Security and Counter-Terrorism Laws. *The Australian Feminist Law Journal* 23: 1–32.

670 Lyon (2008) *op. cit.*

671 van der Ploeg (2005b) *op. cit.*

672 Wickins (2007) *op. cit.*

673 European Commission Joint Research Centre, Institute for Prospective Technology Studies (2005) *op. cit.*

674 Mordini (2008) *op. cit.*

675 Henschke (2007) *op. cit.*

676 Hunter S (2003). A critical analysis of approaches to the concept of social identity in social policy. *Critical Social Policy* 23(3): 322–344.

677 *ibid.*

678 Henschke (2007) *op. cit.*

679 National Consultative Ethics Committee for Health and Life Sciences (2007) *op. cit.*

spheres and relationships.[680,681] For example, they may have different personas at work, or in their personal and family life and these personas may change throughout the individual's life.[682] In addition, a person may wish to keep some degree of separation or distinction between these different personas, for example, individuals may not want their employers to know certain personal information about themselves: their sexual orientation, religious beliefs or political opinions. The ability to portray different personas in different contexts offers a degree of personal freedom and enables an individual to exercise some level of control over the information he/she makes available to particular parties as a means of identification.[683]

It is clear then that biometric identification represents only a very small aspect of our personal identity. It in no way reflects the complex factors and interactions that go to comprise our identity in the more elaborate sense. It is also clear that an individual's personal identity is interconnected with his/her personal privacy. According to Jeffrey H. Reiman, "privacy is a social ritual by means of which an individual's moral title to his existence is conferred", which Reiman deems to be a necessary precondition of personhood.[684] In order to become a person, an individual needs to understand that not alone do his/her choices shape his/her destiny, but also that he/she is exclusively entitled to shape his/her own destiny. Therefore, privacy is an essential requirement for the creation of a person out of a human being because privacy facilitates the understanding that our existence, *i.e.* our thoughts, our body and our actions, are indeed our own,[685] which is important for the attribution of moral responsibilities.

The idea that a given individual has moral title or ownership over his/her existence has important practical, as well as philosophical, implications. Moral ownership implies that each of us, as persons, has the right to control what we do with our own bodies and the right to control when and by whom our body is experienced.[686] It is the possession of a right to privacy that protects and ensures this moral ownership.[687,688] Moreover, this right to privacy is simultaneously interconnected with the notion of an individual's right to bodily integrity and the inviolability of his/her body, which implies that the boundary of the body should not be crossed without the owner's consent.[689,690] Nonetheless, while this boundary is easy to quantify for the physical body, it may become more difficult to define when applied to personal and sensitive information, such as biometrics, derived from an individual's body.[691]

---

680 Nabeth T and Hildebrandt M (2005). *D 2.1: Inventory of topics and clusters*. Future of Identity in the Information Society (FIDIS) Project Deliverable Version 2.0, 57p.
Available online at: http://www.fidis.net/fileadmin/fidis/deliverables/fidis-wp2-del2.1_Inventory_of_topics_and_clusters.pdf, accessed 28 July 2008.

681 Biometric Information Technology Ethics (2005) *op. cit.*

682 Albrecht *et al.* (2003) *op. cit.*

683 Nabeth and Hildebrandt (2005) *op. cit.*

684 Reiman (1984) *op. cit.*

685 *ibid.*

686 *ibid.*

687 Alterman (2003) *op. cit.*

688 Reiman (1984) *op. cit.*

689 van der Ploeg (2005a) *op. cit.*

690 Alterman (2003) *op. cit.*

691 van der Ploeg (2005a) *op. cit.*

The Council recognises the need to establish and/or corroborate the identification of an individual in a globalised world and the many advantages of so doing. However, the method(s) of identification used should in no way be taken to define or categorise a person's identity in a more substantive sense. Indeed, the inappropriate use of bodily information to categorise, stigmatise or discriminate in any way should be resisted strongly. With that in mind, the Council recommends that respect for human dignity should be at the forefront of considerations by policy makers and the biometrics community when designing, implementing and operating biometric technologies and applications.

Informational Privacy

Many of the privacy concerns relating to biometric information can be distilled down to the ability of an individual to retain the control over this information and who has access to it. The philosophical foundations of the right to privacy, previously outlined in this document, would suggest that the loss of this element of control results in a loss of privacy. Moreover, the inability to control information pertaining to us also has negative connotations for the degree of autonomy, dignity and respect shown to us as persons.

An individual's biometric information is an intrinsic element of that person. The Council, therefore, recommends that the right to bodily integrity and respect for privacy should apply not only to an individual's body, but also to any information derived from the body, including his/her biometric information.

Privacy and the Right to Anonymity

Stemming from the concept of an individual's control over information relating to him/her, many people would prefer to keep their biometric (and other personal) information private and confidential and to only make it available to others on their own terms. An individual's right to privacy facilitates this ability to withhold personal information. Therefore, a relationship exists between the notion of privacy and the concept of identification – for example, it is often considered that when an individual's identity is not known he/she has more privacy.[692]

By facilitating the ability to control the availability of information about oneself, the right to privacy necessarily offers the possibility of anonymity. Furthermore, while it is accepted that in many situations an individual can rightfully be expected to identify him/herself, there is also an expectation of a right to anonymity and the freedom (autonomy) to make certain decisions (*e.g.* casting secret ballots in elections) and conduct activities during his/her daily

692 Cavoukian A (2006). *7 Laws of Identity: The Case for Privacy-Embedded Laws of Identity in the Digital Age*. Information and Privacy Commissioner/Ontario, Toronto, 18p. Available online at: http://www.ipc.on.ca/images/Resources/up-7laws_whitepaper.pdf, accessed 7 February 2008.

life without always having to make him/herself known or to make this information known.[693,694] In representing "something you are", biometric modalities enable the ascription of fixed identities to individuals. As a result, the proliferation of biometric technologies could further limit an individual's ability to remain anonymous and therefore maintain his/her privacy in particular circumstances,[695] for example, with regards to political affiliations, religious beliefs or sexual orientation.

However, it is also important to recognise that the ability of biometric modalities to provide stronger authentication of identity has also been used to protect privacy and anonymity. For example, in some clinics in the US, when an individual undergoes an AIDS test he/she can use fingerprint or iris recognition to verify who he/she is, *i.e.* the patient associated with a particular test file, while still remaining anonymous to the medical personnel.[696,697] Moreover in the United Nations (UN) refugee programme on the border of Pakistan and Afghanistan, an anonymous biometric database is used to manage the allocation of aid to the refugees. Each refugee has his/her iris collected and compared against those in the UN database and if no match is found that individual is given his/her aid package. However, if a match is found in the database this indicates that the individual concerned has already received an aid package and he/she is not given a second one.[698]

Collection of the Appropriate Information

Notwithstanding situations where an individual wishes to remain anonymous, when an individual participates in a biometric programme, whether compulsory or voluntary, it is usually done so on the understanding that the information being collected will be used for a specified purpose. For many people, the lack of a definitive, specified purpose underpinning the collection and use of biometric information increases the likelihood of this information being put to other uses (*i.e.* function creep),[699,700] which raises a number of privacy concerns.

To alleviate such concerns it is therefore important that the information collected for a biometric application is limited to that information necessary to identify a given individual participating in the application. Many people believe that any additional personal information that may be collected incidentally during the enrolment or comparison phases should be deleted and not held on to in case it might be deemed useful at some time in the future.[701,702] Of particular concern is the possibility of deriving additional health, medical and sensitive

---

693 Nabeth and Hildebrandt (2005) *op. cit.*

694 Woodward *et al.* (2001) *op. cit.*

695 *ibid.*

696 Harel A (2009). Biometrics, Identification and Practical Ethics. In E Mordini and M Green (eds.) *Identity, Security and Democracy: The Wider Social and Ethical Implications of Automated Systems for Human Recognition.* Volume 49 NATO Science for Peace and Security Series - E: Human and Societal Dynamics, IOS Press, Amsterdam, p.69–84.

697 Biometric Information Technology Ethics (2005) *op. cit.*

698 Most (2004b) *op. cit.*

699 European Commission Joint Research Centre, Institute for Prospective Technology Studies (2005) *op. cit.*

700 Snijder (2007) *op. cit.*

701 National Consultative Ethics Committee for Health and Life Sciences (2007) *op. cit.*

702 Article 29 Data Protection Working Party (2003) *op. cit.*

personal information from certain biometric identifiers, the use of which could have far reaching implications for the individuals involved.[703,704]

Biometric technologies that involve analysis of patterns of blood vasculature, for example, retinal scanning, facial thermography, vein pattern recognition, all have the potential to reveal information about certain conditions, for example, vascular dysfunction, hypertension and pregnancy.[705] Fingerprint patterns have also been shown to reveal information about particular illnesses and conditions. For example, among individuals with Down's Syndrome, Turner's Syndrome and Klinefelter's Syndrome, particular fingerprint patterns predominate.[706,707] In addition, a link has been found between certain fingerprint patterns and incidences of chronic intestinal pseudo-obstruction.[708] The use of liveness detection (*e.g.* pupillary reflex, blood pressure, pulse rate and respiration rate) as part of the recognition process could also provide additional information on a given individual's physiology, medical condition or his/her emotional state.[709] For example, when collecting a live image of an individual's iris his/her pupillary responses could reveal information about that individual's drug and/or alcohol use.[710] Finally, face recognition could also be said to reveal the emotional state of the individual, although in many biometric systems this would be less likely since the system requires users to maintain a neutral pose when interacting with the camera (sensor). However, facial recognition could still indicate additional information about the user, which may not be relevant or necessary for the designated purpose of the system, for example, the individual's race, religion, or culture, which could potentially be used for profiling and categorisation purposes.

In line with the *Data Protection Acts* (1988 and 2003), the Council recommends that biometric systems should only collect that information required to fulfil a prescribed purpose. Since the overarching purpose of biometric systems is to verify or identify a given individual, any additional medical or sensitive personal information collected incidentally, which is not needed for recognition purposes, should be deleted from the system.

Rights of Access and Redress

Since privacy, autonomy and bodily (informational) integrity are related to the control and ownership of personal information, it is generally accepted that every individual should be entitled to know what information about them is being stored, why it is being stored, where it is being stored and who has access to it.[711,712] Privacy and ownership rights are also deemed to

703 Woodward *et al.* (2001) *op. cit.*

704 Mordini and Massari (2008) *op. cit.*

705 Woodward *et al.* (2001) *op. cit.*

706 Mordini (2008) *op. cit.*

707 Woodward *et al.* (2001) *op. cit.*

708 Pulliam TJ and Schuster MM (1995). Congenital markers for chronic intestinal pseudoobstruction. *The American Journal of Gastroenterology* 90(6): 922–926.

709 Mordini and Massari (2008) *op. cit.*

710 Mordini and Ottolini (2007) *op. cit.*

711 Article 29 Data Protection Working Party (2003) *op. cit.*

712 Snijder (2007) *op. cit.*

entitle a given individual to ensure the accuracy of any information that is stored about him/her and enable him/her to redress any errors in that information.[713,714] However, under certain circumstances (*i.e.* in the interest of the common good), for example, where the information is required as part of a criminal investigation, an individual may be prohibited from accessing, reviewing and/or amending information pertaining to him/her.[715] Notwithstanding such restrictions on accessing the information, the Council takes the view that the information stored about an individual should be kept accurate and up to date. It is therefore important that system operators implement some form of review and correction mechanism.

Furthermore, it has been argued that biometric systems and databases should undergo regular audits to ensure that the information is not only correct, but necessary to fulfil the purpose it was collected for.[716,717] Such auditing may help to alleviate concerns relating to the continued storage of biometric and personal information once an individual has left the biometric programme, for example, if he/she has withdrawn his/her consent to participate, if he/she no longer works for a particular company or attends a particular school that had implemented a biometric programme, or even if the individual has died. Given the expected longevity of many national and international biometrics programmes, the issues surrounding the continued storage and use of information related to an individual who has died are likely to arise in relation to these applications.[718]

An individual should have the right to access any collected and/or stored information relating to him/her and to review and amend it where necessary, subject to legal exceptions. Moreover, if an individual no longer wishes to utilise the biometric application or the original purpose of the application has been achieved, then any biometric and other personal information about that person should be deleted from the system.

Function Creep and Interoperability

Granting an individual access to review and amend stored information pertaining to him/her may also help to prevent this information being used for additional, previously non-specified purposes, *i.e.* function creep. However, given the special nature of biometric information, *i.e.* its indelible association with an individual's identity, the possibility of deriving additional information from it, and the potential to use it for categorising individuals, it has been argued that the temptation for function creep is too great to be ignored.[719,720] Developments and efforts towards biometric system interoperability have also been used to bolster this argument

713 OECD, Working Party on Information Security and Privacy (2004) *op. cit.*

714 Data Protection (Amendment) Act 2003.

715 The common good argument is discussed in greater detail in the section entitled 'The Common Good'.

716 Data Protection Commissioner (2007). *Biometrics in Schools, Colleges and other Educational Institutions.* Available online at: http://www.dataprotection.ie/docs/Biometrics_in_Schools_Colleges_and_other_Educational_Institu/409.htm, accessed 5 November 2007.

717 International Biometric Group (2007a) *op. cit.*

718 For example, the retention period for the US-VISIT programme is set at 75 years.

719 Woodward *et al.* (2001) *op. cit.*

720 Davies SG (1994). Touching Big Brother: How biometric technology will fuse flesh and machine. *Information Technology & People* 7(4): 38–47.

of the inevitability of function creep.[721,722] While the current lack of absolute interoperability between all systems is often quoted as a means of allaying some privacy concerns, this reasoning is no longer considered sufficient since standards for interoperability have been, and are being, developed.[723] The more agencies and organisations that have access to an individual's biometric information, the greater the likelihood that this information will be used for another purpose beyond that for which it was originally collected.[724,725] The evolution and extension of the use of the social security number in the US, from its original introduction for administrative purposes at the federal and state level to its automatic requirement as a form of identification for a host of routine transactions, is often given as an example of the ease with which function creep can occur.[726,727,728] To help avoid such occurrences, it is frequently suggested that government and commercial or other private databases should not share information with each other.[729,730,731]

The use of the Personal Public Service (PPS) number, a unique reference number given to all Irish citizens and all individuals working in Ireland, has expanded significantly since its initial inception as the Revenue and Social Insurance number, which was an individual identifier for transactions between a given individual, the Department of Social and Family Affairs (DSFA) and the Revenue Commissioners.[732] The PPS number is now required for an individual to gain access to a wide range of services including all social welfare services and benefits, public health services and the free travel pass, and it has been suggested that its use will be further expanded in the future.[733] However, concerns have been raised, by both the Office of the Data Protection Commissioner and the DSFA, about the potential for the PPS number to be used as a "unique identifier for a multitude of unspecified purposes" by public and private bodies.[734] Legislation, in the form of the *Social Welfare Acts*,[735] regulates which bodies can legitimately use a given

721 van der Ploeg (2005b) *op. cit.*

722 Harel (2009) *op. cit.*

723 Alterman (2003) *op. cit.*

724 Woodward *et al.* (2001) *op. cit.*

725 van der Ploeg (2005b) *op. cit.*

726 Etzioni (1999) *op. cit.*

727 Woodward *et al.* (2001) *op. cit.*

728 The Canadian equivalent of the social security number has also become the default identity requirement in numerous transactions in Canada.

729 Snijder (2007) *op. cit.*

730 National Consultative Ethics Committee for Health and Life Sciences (2007) *op. cit.*

731 Crews CW Jr (2002). Human Bar Code: Monitoring Biometric Technologies in a Free Society. *Policy Analysis* 452: 1–20. Available online at: http://www.cato.org/pubs/pas/pa452.pdf, accessed 30 May 2008.

732 The PPS number was originally introduced as the Revenue and Social Insurance (RSI) number in 1993.

733 Department of Social and Family Affairs (2008). *Personal Public Service Number.* SW 100, Department of Social and Family Affairs, Sligo, 4p. Available online at: http://www.welfare.ie/EN/Publications/SW100/Documents/sw100.pdf, accessed 14 August 2008.

734 Data Protection Commissioner (2008b). *Data Protection in the Department of Social & Family Affairs. Report by the Data Protection Commissioner.* Data Protection Commissioner, Portarlington, 37p.
Available online at: http://www.welfare.ie/EN/Topics/Documents/ODPCReport.pdf, accessed 1 August 2008.

735 The Social Welfare (Consolidation) Act 1993 as amended by the Social Welfare Acts (1994-2007), which includes the Social Welfare (Consolidation) Act 2005 and the Social Welfare and Pensions Act 2007.

individual's PPS number and associated information (*i.e.* his/her public service identity)[736] in any dealings with that individual. However, it has been acknowledged that the DSFA's registry of bodies authorised to utilise PPS numbers needs to be managed and monitored more tightly, to ensure it complies with social welfare and data protection legislation.[737]

The Irish government is proposing to introduce a public service card, which could include an individual's public service identity as well as his/her photograph and signature,[738] though the final decision on what exact information will be kept on the card is yet to be made.[739] It has been proposed that this public service card will replace other cards currently used for accessing services in social welfare, revenue, health, education, agriculture and local government.[740] However, while it is hoped that this card will increase efficiency and security when availing of government services, concerns have been raised that this card could potentially become a default national ID card.[741,742] It is proposed that the initial roll out of the card will be for those individuals eligible for a free travel pass, for example, individuals over 65 years of age. However, it should be noted that, while the use of the card itself will be optional, individuals will not be able to participate in the Free Travel Scheme without it.[743]

One of the main examples of biometric system interoperability at the global scale relates to the ICAO recommendations for the use of facial recognition images in machine readable travel documents. To ensure global compatibility with border management and immigration programmes, the ICAO recommended that raw facial images be stored on such travel documents.[744] Doubts have been expressed about the ability to control the use of, and access to, the biometric and associated information once it has served its initial purpose of authenticating an individual's identity.[745,746,747] For many the potential availability of such biometric information to a host of third parties represents a further invasion of privacy and an erosion of autonomy.[748,749] It has even been argued that increasing interoperability between

---

736 An individual's PPS number when combined with other information such as, his/her name (and any previous surnames), date of birth, mother's former surname, sex, nationality and address represents his/her public service identity.

737 Data Protection Commissioner (2008b) *op. cit.*

738 Social Welfare and Pensions Act (2007).

739 Data Protection Commissioner (2008b) *op. cit.*

740 O'Brien C (2008). Photo ID cards may face cash constraint delays. *The Irish Times* 6 August 2008. Available online at: http://www.irishtimes.com/newspaper/ireland/2008/0806/1217923985190.html, accessed 14 August 2008.

741 *ibid.*

742 Coyle C (2008). Irish bus pass is "identity card by stealth". *The Sunday Times* 10 August 2008. Available online at: http://www.timesonline.co.uk/tol/news/world/ireland/article4493788.ece, accessed 14 August 2008.

743 Data Protection Commissioner (2008b) *op. cit.*

744 ICAO TAG (2004) *op. cit.*

745 De Hert P, Schreurs W and Brouwer E (2007). Machine readable identity documents with biometric data in the EU – part III. *Keesing Journal of Documents & Identity* 23: 27–32.

746 van der Ploeg (2005a) *op. cit.*

747 European Commission Joint Research Centre, Institute for Prospective Technology Studies (2005) *op. cit.*

748 *ibid.*

749 Cavoukian (1999) *op. cit.*

biometric systems and programmes brings us further down the slippery slope to a "Big Brother" society.[750,751,752]

Despite these concerns, interoperability also offers a number of positive characteristics – such as helping to reduce system operation costs, limiting lock-in to particular vendors, algorithms and matching technology, as well as increased convenience for the system user, for example, he/she could use the same travel document no matter where he/she was travelling to. Moreover, interoperability is said to facilitate national and international policies relating to security and public safety, for example, counterterrorism, identity theft, organised crime, and illegal migration, which enables states to better protect the rights of their citizens.[753,754] It has even been suggested that preventing the sharing of information about criminals or those who would attempt to do harm between countries would be unethical.[755] While acknowledging that increased interoperability could potentially facilitate function creep, the Council also recognises that interoperability, if employed correctly, can improve the functionality of particular biometric programmes.

The implementation of basic management measures to control access to, and the use of, a given database can also help to alleviate concerns relating to function creep and system interoperability. These management measures could involve establishing a hierarchy of access to the information such that different administrators can only access levels of biometric and personal information commensurate with their position and necessary to conduct their particular job.[756,757,758] In addition, staff who would be using the database could be trained in the appropriate use of such information. Furthermore, limitations could also be placed on the possibility of downloading or copying information from a database, unless it is absolutely necessary, thereby reducing the risk of such information being shared inappropriately or being lost or stolen. Where information is to be, legitimately, shared with third parties, protocols could be established, whereby these individuals gain access only to the specific information they require to complete their particular task. There have been numerous examples, in recent years, of data losses from different government and other agency databases, which could possibly have been prevented, or at least made less likely, if adequate security and information management procedures had been in place.

---

750 Woodward (2001) *op. cit.*

751 Lodge (2007) *op. cit.*

752 Lyon (2008) *op. cit.*

753 European Commission Joint Research Centre, Institute for Prospective Technology Studies (2005) *op. cit.*

754 Snijder (2007) *op. cit.*

755 National Science and Technology Council (NSTC) Subcommittee on Biometrics (2006l). *International Conference on Biometrics and Ethics, Conference Highlights.* NSTC, Washington, 3p. Available online at: http://www.biometrics.gov/NSTC/Conference_on_Biometrics_and_%20Ethics_2006_highlights.pdf, accessed 28 July 2009.

756 Data Protection Commissioner (2007) *op. cit.*

757 Collins J (2007). Data insecurities. *The Irish Times* 23 November 2007. Available online at: http://www.irishtimes.com/newspaper/finance/2007/1123/1195682113657.html, accessed 4 February 2008.

758 It is important to note that the log in and management systems established to control access to a given database could, themselves, be biometric systems, in addition to or in place of user PINs and passwords.

It has been reported that over 120 data storage devices, including laptops and portable memory devices have been lost or stolen from Irish government departments since 2002,[759] and 16 laptops belonging to the Comptroller and Auditor General have been stolen since 1999.[760] Many of these devices contained personal and sensitive information of state employees and the general public, which in some cases was not stored in an encrypted or otherwise appropriately secure format.[761] While incidences of theft may not be entirely preventable, security breaches and inappropriate use of personal information pertaining to the general public can still occur within government departments and other organisations. For example, it was revealed that personal information held by the Department of Social and Family Affairs (DSFA), which related to an individual who had won the lottery, was accessed by over 100 staff members of the DSFA, only 34 of whom had a legitimate reason to access the information.[762]

The Council is of the opinion that, in order to respect and uphold an individual's privacy and confidentiality, biometric applications should utilise only information required to meet a clear, limited and specified purpose. Therefore, any subsequent attempts to use the information for another purpose or to share it with third parties without the knowledge and consent of the individual should be prohibited.

In addition, the Council recommends that appropriate information and access management procedures should be established for all biometric applications to ensure that:

- system operators and system providers are properly trained with regard to their obligations to respect and protect the information;
- system operators and system providers can access only the information they require to conduct their job.

---

759 Quinlan R (2008). More than 120 data-storage devices "lost" by government staff. *The Sunday Independent* 27 April 2008. Available online at: http://www.independent.ie/national-news/more-than-120-datastorage-devices-lost-by-government-staff-1360451.html, accessed 27 April 2008.

760 McGee H (2008). 16 office laptops stolen since 1999, says comptroller. *The Irish Times* 2 August 2008. Available online at: http://www.irishtimes.com/newspaper/ireland/2008/0802/1217368897558.html, accessed 2 August 2008.

761 Taylor C (2008). Thousands of social welfare details on stolen laptop. *The Irish Times* 11 August 2008. Available online at: http://www.irishtimes.com/newspaper/breaking/2008/0811/breaking25.htm, accessed 11 August 2008.

762 Irish Council for Civil Liberties (2008). *Safeguards Essential in Piloting of Public Service Card Says the ICCL.* Press release from the Irish Council for Civil Liberties, 5 August 2008.

Amalgamating Information and Profiling

The way in which biometric and other personal data that have been collected are compared, stored, and possibly linked to other information about an individual is seen as problematic by many people.[763,764] Understandably, the amalgamation of such information raises the spectre of a person being placed under surveillance and tracked, and a detailed profile being created, resulting in that individual being categorised according to his/her behaviour and activities. Nonetheless, since information derived from an individual's body is considered to be part of the embodied person, issues relating to bodily and, therefore, informational integrity should still apply even when the information is stored in a database.[765] Therefore, it has been argued that database analysis and data mining practices involving biometric (and other personal information) are equivalent to body searches.[766]

The Pakistani government has developed a highly integrated database of approximately 80 million citizens as part of its National Identification Program, which utilises both face and fingerprint recognition systems.[767] This system has been introduced to overcome the problems of incomplete and inaccurate records with the old identity card system. All civil and financial applications now use the National Identity of a citizen as a unique identifier, with the database accessible as part of a citizen verification service to government agencies, financial institutions and telecommunications companies. In addition, the database provides huge data mining capabilities, including the ability to link each individual to other members of his/her immediate and extended family.[768]

Concerns relating to profiling and the amalgamation of information may be exacerbated by the fact that a lot of personal information can already be garnered legally from our everyday activities, which can provide information about a particular individual's activities, interests and background. For example, when performing internet searches the IP (Internet Protocol) address (the unique address of the computer on the internet) and the search term are stored, details of the phone numbers a person calls can be stored, banking and credit card transactions are tracked, library records are recorded and even a supermarket/shop loyalty card can keep track of an individual's purchases.[769] It is also possible to pinpoint an individual's location by tracking his/her mobile (cellular) phone. In addition, given the ubiquitous presence of CCTV (closed circuit television) cameras and other surveillance equipment throughout our cities and towns, many individuals are accustomed to images and footage of themselves being captured. For

---

763 van der Ploeg (2005b) *op. cit.*

764 Alterman (2003) *op. cit.*

765 van der Ploeg (2005a) *op. cit.*

766 *ibid.*

767 Mobin UY (2007). *Pakistan's National Biometric Identification Program.* Presentation at the Biometrics Exhibition and Conference 2007, 17–19 October 2007, Westminster, London.

768 *ibid.*

769 The Irish Times (2007). Surveillance Nation. *The Irish Times* 4 August 2007.
Available online at: http://www.irishtimes.com/newspaper/newsfeatures/2007/0804/1186123297026.html, accessed 9 November 2007.

example, in the UK, where there are an estimated 4.2 million surveillance cameras in operation, the average citizen may be recorded on over 300 cameras each day.[770]

Moreover, based on the interconnection and linkability of the biometrically controlled transactions individuals make, and the systems they use, the processes of profiling, tracking and categorising people are expected to become more relevant as biometric systems and applications become more widespread and their interoperability improves.[771,772] Ostensibly such profiling is often conducted with the intention of improving public safety and/or national security. However, it is considered important not to disproportionately target certain groups within society as this could lead to the erosion of the trust model between individuals and the state if people feel they are being treated unfairly or discriminated against, for example, through racial profiling. Nonetheless, critics of these practices consider it highly probable, if not inevitable, that profiling and social categorisation will focus predominantly on particular groups within society.[773,774] Consequently, there have been calls for increased transparency and openness in the management and implementation of such profiling measures,[775] and the establishment of appropriate safeguards to minimise the risk of abuse in these circumstances.[776,777]

By combining different biometric modalities through processes such as data mining, it may be possible to generate profiles about individuals, which could classify them as potentially risky or suspicious and, therefore, in need of further investigation. Systems are already available to conduct certain background checks, though not necessarily involving biometrics, on an individual. For example, passenger name records relating to all individuals travelling to the US and advance passenger information for individuals from certain countries, including Ireland, travelling to Spain are used in background checks. The inclusion of biometrics could potentially increase the confidence in such background identifications, through the provision of stronger identity authentication. Any profiles generated could then be used to attempt to predict an individual's behaviour and activities. Many regard this form of pre-emptive surveillance as a useful tool in deterring criminals and reducing crime rates.[778]

The Cogito system developed by an Israeli company, Suspect Detection Systems, enables the automated screening of airline passengers.[779] With this system the passenger sits inside a booth, scans his/her passport or identity card and places his/her hand on the sensor. The computer then asks a series of 15 to 20 questions, while

770 House of Commons Home Affairs Committee (2008). *A Surveillance Society? HC-I, Fifth Report of Session 2007-08. Volume I: Report, together with formal minutes.* The Stationery Office Limited, London, 117p.

771 van der Ploeg (2005b) *op. cit.*

772 Commission de l'éthique de la science et de la technologie (2008). *In Search of Balance: An Ethical Look at New Surveillance and Monitoring Technologies for Security Purposes.* Commission de l'éthique de la science et de la technologie, Quebec, 73p.

773 Lyon (2008) *op. cit.*

774 van der Ploeg (2005b) *op. cit.*

775 Snijder (2007) *op. cit.*

776 van der Ploeg (2005b) *op. cit.*

777 International Biometric Group (2007a) *op. cit.*

778 Commission de l'éthique de la science et de la technologie (2008) *op. cit.*

779 Marks P (2007). Can a government remotely detect terrorists thoughts? *New Scientist* 195(2616): 24–25.

the sensor measures the individual's blood pressure, pulse and perspiration rate. The individual's bodily responses are compared to those of other (innocent) people who were asked the same questions previously to assess if that individual represents a threat (*i.e.* displays some degree of hostile intent). Based on this comparison a decision is made on whether or not the individual should be taken for further questioning. Advocates of this system argue that the use of technology to screen passengers in this way circumvents issues associated with subjective (*e.g.* racial) profiling. However, it should be noted that, because it is not practical to screen all passengers in this way,[780] some level of pre-selection and profiling would have to be conducted in order to choose potential participants.[781] In addition, while this technology is considered promising, some system accuracy challenges, specifically false matches and false non-matches, need to be overcome.[782]

Another technology being implemented to screen passengers is based on combining behavioural analysis with biometrics. The aim of the technology is to implement a non-invasive, remote sensing system to identify hostile intent through the analysis of an individual's facial expressions, gait, blood pressure, pulse and perspiration rates, breathing rate and skin temperature. This would enable thorough screening of all passengers without slowing their movements through the airport. The Future Attribute Screening Technologies (FAST) programme,[783] which utilises such technology, is still undergoing trials in the US and while initial results have been promising the issues of system accuracy still need to be resolved.[784]

While limited profiling already occurs for security reasons and for tackling crime, questions have been raised regarding both the success and justification of such measures.[785,786] For example, it has been argued that the use of such profiling techniques has been successful in preventing potential terrorist attacks.[787] However, this is not the consensus view and it is often pointed out that the presence of biometric security measures is not a foolproof measure against terrorist attacks in the future, for example, a number of the terrorists involved in the attacks on September 11th 2001 had travelled to the US using their own passports, with valid visas.

780 Processing a single individual using the Cogito system takes approximately 5 minutes.

781 Marks (2007) *op. cit.*

782 For example, the company are looking for a 90% positive match rate (with 10% false negatives) and a 4% false positive match rate. See Wall Street journal article (Karp J and Meckler L [2006]. Which Travelers Have "Hostile Intent"? Biometric Device May Have the Answer. *The Wall Street Journal* 14 August 2006. Available online at: http://online.wsj.com/public/article/SB115551793796934752-2hgveyRtDDtssKozVPmg6RAAa_w_20070813.html?mod=tff_main_tff_top, accessed 5 December 2008).

783 This programme was previously entitled Project Hostile Intent.

784 Marks (2007) *op. cit.*

785 Bigo D, Carrera S, Guild E and Walker RBJ (2007). *The Changing Landscape of European Liberty and Security: Mid-Term Report on the Results of the CHALLENGE Project.* Research Paper No. 4, 46p. Available online at: http://www.libertysecurity.org/article1357.html, accessed 9 July 2008.

786 Crews (2002) *op. cit.*

787 Mahony H (2007). Database of passenger flight details proposed. *The Irish Times* 7 November 2007. Available online at: http://www.irishtimes.com/newspaper/world/2007/1107/1194222776758.html, accessed 8 November 2007.

Nonetheless, opponents of such profiling practices have questioned their legitimacy through suggestions that individuals might be labelled as suspicious on the basis of very limited amounts of information or insufficient evidence. For these critics the development of biometric standards, increased system interoperability and increased database interconnectivity will facilitate even more pervasive surveillance, categorisation and profiling of individuals.[788,789] Some would consider this outcome as unlikely and unrealistic, given the acknowledgement of human rights and civil liberties within society in addition to the presence of legislation and regulations.[790] Nonetheless, these issues and concerns are seen as integral to the debate on biometric technologies and their associated applications.

The Council believes that it is essential that profiling measures do not target particular groups within society unfairly or disproportionately. In addition, where an individual is profiled, this should be done in an appropriate manner based on valid reasoning and evidence, and in accordance with due process to ensure his/her rights and civil liberties are respected and upheld.

Autonomy

Informed Consent

The previous discussions have emphasised the inherent importance of an individual's body to his/her concept of privacy and identity. Integral to these concepts is the notion of ownership of the body and, necessarily, any personal and biometric information derived from, or relating to, the body or the person. Questions regarding who maintains control of this information and the access to it, relate not only to an individual's privacy and bodily integrity, but also his/her autonomy. Autonomy represents an individual's ability to make decisions or take actions based on his/her own convictions and free from external influences. In general, an individual's right to autonomy is recognised and respected, provided the decisions of the individual do not result in the harming of others.[791] This view of autonomy is encapsulated and elucidated in John Stuart Mill's "liberty principle", which states that "the only part of the conduct of any one, for which he is amenable to society, is that which concerns others. In the part which merely concerns himself, his independence is, of right, absolute. Over himself, over his body and mind, the individual is sovereign".[792]

An integral component of exercising autonomy is the concept of informed consent, with the individual's choice being based on all the details relevant to making the decision. In the case of biometric applications, such details could include what personal information (biometric or otherwise) will be collected as part of the application, for what purpose this information is

788 van der Ploeg (2005b) *op. cit.*

789 Woodward (2001) *op. cit.*

790 European Commission Joint Research Centre, Institute for Prospective Technology Studies (2005) *op. cit.*

791 Irish Council for Bioethics (2007). *Is It Time For Advance Healthcare Directives? Opinion.* Irish Council for Bioethics, Dublin, 98p. Available online at: http://www.bioethics.ie/uploads/docs/Advance_Directives_HighRes.pdf, accessed 17 November 2008.

792 Mill JS (1863). *On Liberty.* 2nd edn. Ticknor and Fields, Boston, 223p.

being collected, how the information will be collected, how and where this information will be stored (*e.g.* as a template and/or a raw image, encrypted or un-encrypted, *etc.*), who will have access to the information, how long it will be stored for, whether or not he/she will be able to see the stored information and amend it or remove it if necessary, as well as the benefits and possible risks of participating or not in the biometric programme.

Therefore, in order for an individual to provide his/her informed consent, it is important that he/she understands the purpose and the implications of the proposed system and the potential consequences of his/her own decision to participate or not.[793,794,795] While in many cases an adult could be considered to have a sufficient grasp of the information at hand before exercising his/her autonomy, concerns have been raised in relation to children providing consent to participate in biometric applications in schools.[796] In some instances, children as young as four or five years of age may be enrolled into these systems and some concerns have been raised regarding the ability of these children to fully understand the implications of taking part in the biometric system or even how to interact with the system correctly.[797,798] Similar concerns have also been raised in relation to other vulnerable groups within society, for example, the elderly, those with degenerative mental conditions, or those with learning difficulties or mental disabilities.[799,800] It is thus considered important for appropriate safeguards and mechanisms to be put in place to ensure these individuals are both protected and not excluded or disadvantaged.[801,802]

In the case of children, one safeguard that is usually implemented is the requirement that the parent or guardian of the child also provides his/her consent. Questions have arisen regarding at what age a child would no longer need the added consent of a parent or guardian. For example, in Ireland, the Data Protection Commissioner has implemented non-binding guidelines on the use of biometrics in schools, which state, in relation to consent, that a student aged 18 or older should give consent him/herself, a student aged between 12 and 17 should give consent him/herself in conjunction with his/her parent or guardian, and for students younger than 12 only the consent of the parent or guardian would be necessary.[803] However, it should be noted that guidance on the use of biometrics in schools in the UK states that, under the *Data Protection Act 1998*, though it is necessary to inform the student and his/her parent or guardian about the biometric programme it is not always necessary to obtain the consent of

793 Data Protection Commissioner (2007) *op. cit.*

794 European Commission Joint Research Centre, Institute for Prospective Technology Studies (2005) *op. cit.*

795 National Consultative Ethics Committee for Health and Life Sciences (2007) *op. cit.*

796 Data Protection Commissioner (2007) *op. cit.*

797 *ibid.*

798 Futurelab (2007). Should we allow Big Brother in schools? *Vision* 4: 1–4. Available online at: http://www.futurelab.org.uk/resources/documents/vision/VISION_04.pdf, accessed 16 May 2008.

799 Wickins (2007) *op. cit.*

800 NSTC Subcommittee on Biometrics (2006d) *op. cit.*

801 European Commission Joint Research Centre, Institute for Prospective Technology Studies (2005) *op. cit.*

802 Wickins (2007) *op. cit.*

803 Data Protection Commissioner (2007) *op. cit.*

the parent or guardian.[804,805] Only in cases where the student is considered not to understand what is involved need the consent of a parent or guardian be obtained.[806] However, concerns have been raised in relation to the lack of the need for parental consent and in relation to the lack of clarity in the guidelines in relation to this issue.[807]

The widespread concern regarding the collection of biometric information from children in schools is also manifest in the perceived habituation or "softening up" of children to provide their biometric and personal information without fully understanding the implications of this.[808,809] It could be argued that such children would develop less regard for their personal information and privacy and would, consequently, be more willing to accept a greater level of intrusion as standard in the future.[810] Some opponents to the collection of biometric information from children have stated that such activities are unjustified, disproportionate and an unnecessary invasion of the child's right to privacy as outlined in the *UN Convention on the Rights of the Child* (1990).[811]

In order to make the decision whether or not to participate in a biometric programme an individual should be fully and accurately informed and should understand all the issues and implications relating to the provision of his/her information. The Council considers that the issue of user understanding is of particular importance for biometric applications that will be used by potentially vulnerable groups (*e.g.* the elderly, the very young or those with mental and/or learning disabilities). Where such individuals are deemed competent and aware of the consequences of their decision, this decision should be respected. However, if the person is not considered competent, decisions regarding his/her participation should be made by his/her parent or legal guardian. In the case of biometric applications involving children (*i.e.* individuals under 18 years of age), the assent of the child should be sought as well as the consent of his/her parent or legal guardian.

Covert Collection of Biometric Information

Given the advances in surveillance technologies and the potential for remote and distant sensing of certain biometrics, some personal and biometric information could potentially be acquired without an individual's knowledge or express consent. A common example of this involves surveillance cameras, which collect images and footage of people without their

804 Information Commissioner's Office (2008). *The use of biometrics in schools.* V1.1 August 2008. Information Commissioner's Office, Cheshire, UK, 3p. Available online at: http://www.ico.gov.uk/upload/documents/library/data_protection/detailed_specialist_guides/fingerprinting_final_view_v1.11.pdf, accessed 15 September 2008.

805 Becta (2007). *Becta guidance on biometric technologies in schools.* Version 1 July 2007. Becta, Coventry, UK, 10p. Available online at: http://schools.becta.org.uk/upload-dir/downloads/becta_guidance_on_biometric_technologies_in_schools.doc, accessed 5 November 2007.

806 *ibid.*

807 Action on Rights for Children (ARCH) (2007). *Child Tracking: Biometrics in Schools & Mobile Location Devices.* Available online at: http://www.arch-ed.org/issues/Tracking%20devices/final_report_on_child_tracking.htm, accessed 16 May 2008.

808 Article 29 Working Party (2003) *op. cit.*

809 Data Protection Commissioner (2008) *op. cit.*

810 Action on Rights for Children (ARCH) (2007) *op. cit.*

811 Further information about the UN Convention on the Rights of the Child (1990) is available here: http://www.unicef.org/crc/

express consent for the purposes of crime prevention and investigation. Nonetheless, the Council takes the view that there are only a limited number of scenarios where the covert collection of biometric information may be justified.

In 2001 a facial recognition system was used to covertly screen the audience attending the Super Bowl in the US for comparison against a watch list of known criminals. When news of these surveillance measures came to light there was widespread criticism of the system and suggestions that the authorities were checking the identities of those who attended the game, without their knowledge.[812,813] In actual fact, those operating the facial recognition surveillance system did not know and, more importantly, did not need to know the identities of the individuals in the crowd, *i.e.* they remained anonymous. The system was designed to recognise those individuals who were already on the watch list and only those people who looked like someone on the watch list might have been required to confirm their identity.[814] It is difficult to know if the criticism of the biometric system in this instance would have been as harsh if there had been greater transparency from the system operators in advance of its implementation.

In terms of biometric modalities, currently facial and, to a lesser degree, gait biometrics lend themselves to distance collection; however, for the majority of biometric identifiers, for example, fingerprint, palm print, hand geometry, ear geometry and hand vasculature, covert collection is still difficult, since some level of user cooperation is generally required, and/or remote collection is not yet fully feasible. While consent to collect biometric information may not be sought, it is generally a requirement to notify people that they could be under surveillance, for example, with a notice proclaiming that CCTV cameras are in operation in that area. The provision of information in relation to such surveillance programmes can be an important aspect of increasing public awareness and understanding of the programme in operation and its purpose. In addition, providing information to the public may also help to assuage privacy and civil liberties concerns and, ultimately, increase acceptance of such measures.[815,816]

In addition, recent developments in video surveillance technology may also help to alleviate some concerns regarding the covert collection of biometric information. A company (3VR) in the US has developed an image scrambling algorithm to be used in conjunction with its new facial recognition software. While the facial recognition system is used to identify known suspects (and individuals from watch lists) in the surveillance footage, the image scrambling algorithm is used to blur the faces and bodies of those individuals also in the video footage who are not of interest to the system operators, *i.e.* innocent people. The blurred images are also encrypted as a further security and privacy protective measure.[817] Nonetheless, despite these developments, some privacy advocates still question the need for CCTV to record

812 Bowyer (2004) *op. cit.*

813 Woodward (2001) *op. cit.*

814 Bowyer (2004) *op. cit.*

815 Woodward (2001) *op. cit.*

816 Commission de l'éthique de la science et de la technologie (2008) *op. cit.*

817 New Scientist (2009). Encrypted CCTV protects the innocent. *New Scientist* 2717: 19.

surveillance footage constantly, which entails collecting footage of innocent people, as opposed to only recording when something suspect is detected.[818]

Where biometric information is to be collected without an individual's cooperation, the Council considers that, subject to legal exceptions, system operators have an obligation to notify the potential participants (whether willing or unwilling) that the collection of biometric information is ongoing in that area. Moreover, system operators should also provide some explanation as to why the biometric information is being collected and who will have access to it.

The Ability to Opt Out

Consent also implies that an individual should be able to make a voluntary choice regarding his/her participation in a biometric application.[819] There may be situations where an individual does not wish to participate in a biometric application, *i.e.* he/she opts out. An individual can make his/her decision for a variety of personal, cultural or religious reasons.[820] If an individual chooses not to participate in a particular biometric programme he/she should not be disadvantaged or discriminated against and alternative non-biometric means of accessing the same services/entitlements should be provided.[821,822] Moreover, it is considered important not to discriminate against users of non-biometric systems by downgrading or neglecting such systems as a means of encouraging or coercing people to use a related biometric system instead.[823] Individuals should not feel under pressure or compelled to enrol in a biometric programme because their work colleagues are willing to do so or because non-participation could result in some level of stigmatisation.[824,825]

In many cases, biometric systems are completely voluntary and are promoted on the basis of the benefit these systems can provide to those individuals willing to enrol. For example, the INSPASS system, based on hand geometry, was instigated to reduce the need for regular travellers to go through the entire immigration process every time they left or re-entered the US.[826] In addition, in the case of Project IRIS, a border management and immigration programme in the UK, the system is voluntary, and those individuals who sign up, intend to benefit from the fast, efficient and secure passage through immigration, *i.e.* they sign up to beat the queues.

While an individual's biometric or other personal information should, ideally, only be collected with his/her consent, it may be possible to override the requirement for consent under certain

818 *ibid.*

819 Alterman (2003) *op. cit.*

820 For example, some Christians object to using biometrics on the grounds that it relates to the "Mark of the Beast" as referred to in the book of Revelation in the Bible (Woodward *et al.* (2001) *op. cit.*).

821 Data Protection Commissioner (2007) *op. cit.*

822 Wickins (2007) *op. cit.*

823 Harel (2009) *op. cit.*

824 Data Protection Commissioner (2008a) *op. cit.*

825 Davies (1994) *op. cit.*

826 The INSPASS programme was disbanded following the events of 11th September 2001.

circumstances. It may not always be possible for an individual to opt out of a biometric application, for example, numerous biometric initiatives relating to international travel and migration are compulsory.[827] The rationale provided for the compulsory nature of these programmes relates to acting for the benefit of the common good.

Notwithstanding certain compulsory biometric applications, the Council recommends that an individual should be entitled to exercise his/her autonomy freely and without any external influences when deciding whether or not to enrol in a given application. The Council considers it important that non-biometric alternative systems should be made available, where practicable, for those individuals who do not want to use the biometric system, and individuals should not be disenfranchised or discriminated against by choosing not to participate in a given biometric programme.

The Common Good

Limiting Individual Rights for the Common Good

While an individual's personal autonomy should ideally be respected and upheld, this right also has to be balanced against the needs of society overall, *i.e.* the common good. However, in the context of biometric applications, where deference to the common good prevails over autonomy, this could, potentially, result in the diminishment of an individual's privacy.

In political philosophy, libertarians argue that the state and its institutions are not entitled to impinge on the rights of citizens but are, rather, obliged to protect and uphold those rights. From this libertarian point of view, individual rights restrict the actions of other people, in effect, they draw a line around a person, which no one else is entitled to cross.[828] Indeed, it has been argued that:

> Each person possesses an inviolability founded on justice that even the welfare of society as a whole cannot override. For this reason justice denies that the loss of freedom for some is made right by a greater good shared by others. It does not allow that the sacrifices imposed on a few are outweighed by the larger sum of advantages enjoyed by many.[829]

Using this perspective, it would appear that governments are not entitled to encroach individual rights to support a concept of a common good. In fact, libertarians query the validity of the concept altogether, when they argue that only individuals have rights and that society

827 For example, participation in the Eurodac system in the EU is compulsory for all asylum seekers.

828 Berlin I (1969). *Four Essays on Liberty*. Oxford University Press, Oxford, 32p.
Available online at: http://www.nyu.edu/projects/nissenbaum/papers/twoconcepts.pdf, accessed 20 November 2008.

829 Rawls J (1994). *Ethics in the Public Domain*. Oxford Clarendon Press, Oxford, 250p.

is merely a metaphysical theory.[830] Therefore, the compulsory use of biometrics to protect the common good might, from this standpoint, be seen to be morally questionable.[831]

A number of schools of thought, including communitarianism, Rawlsian liberalism[832] and socialism[833], argue in favour of the need to balance individual rights and interests with that of society as a whole, and that autonomous beings are shaped by the culture and values of their communities. For example, the moral philosopher Alasdair MacIntyre observes that:

> ... we all approach our own circumstances as bearers of a particular social identity. I am someone's son or daughter, someone else's cousin or uncle; I am a citizen of this or that city, a member of this or that guild or profession; I belong to this clan, that tribe, this nation. Hence what is good for me had to be good for one who inhabits these roles. As such, I inherit from the past of my family, my city, my tribe, my nation, a variety of debts, inheritances, rightful expectations and obligations.[834]

In relation to biometric technologies specifically, Amitai Etzioni, a communitarian scholar, states that:

> American society incurs high costs – social, economic, and other kinds – because of its inability to identify many hundreds of thousands of violent criminals, white-collar criminals, welfare and credit card cheats, parents who do not pay child support, and illegal immigrants. If individuals could be properly identified, public safety would be significantly enhanced and social and economic costs would be reduced significantly. [...] We must hence ask: Do the benefits to public safety and other public goals of ID cards or biometrics outweigh the cost to privacy?[835]

If we are to adopt the communitarian approach, we must observe that an individual's rights are not absolute and cannot be upheld to their fullest extent in every situation. Essentially, upholding the rights of one individual in a given situation could have a negative impact on the rights of other individuals.[836] This is indicative of the interconnectedness of all individuals within society and, therefore, some form of balance needs to be struck between upholding the rights of a given individual on the one hand and upholding the rights of society at large on the other.[837,838]

---

830 Narveson J (2002). Collective Responsibility. *The Journal of Ethics* 6(2): 179–198.

831 Weber K (2006). *The Next Step: Privacy Invasion by Biometrics and ICT Implants.* Presentation given at the Zif Workshop on Privacy, February 2006. Available online at: http://www.acm.org/ubiquity/views/pf/v7i45_weber.pdf, accessed 20 November 2008.

832 Rawls J (1993). *Political Liberalism.* Cambridge University Press, New York, 464p.

833 King, P (1996). *Socialism and the Common Good. New Fabian Essays.* Taylor and Francis Inc., London, 336p.

834 MacIntyre A (1994). The Concept of a Tradition. In M Daly (ed.) *Communitarianism: A New Public Ethics.* Wadsworth Publishing Company, California, p.123–126.

835 Etzioni (1999) *op. cit.*

836 Clarke (1984) *op. cit.*

837 Commission de l'éthique de la science et de la technologie (2008) *op. cit.*

838 Etzioni (1999) *op. cit.*

This need for balance was recognised by John Stuart Mill when he formulated his "liberty principle". While Mill argued that an individual should have the freedom to act independently and without the interference of others, he also recognised that this freedom could apply only to those actions or decisions taken by an individual that did not affect anyone else.[839] Mill considered that "the only purpose for which power can be rightfully exercised over any member of a civilised community, against his will, is to prevent harm to others".[840] The Council takes the view that Mill's rationale can be applied to the way in which the state has the power to intervene in particular situations and override the rights of an individual to protect society at large, in limited circumstances.[841] For example, the state may place an individual suffering from a highly contagious disease into quarantine to reduce the risk of this disease spreading further in the population.[842,843] The reason for the intervention of the state in such situations derives from its duty to uphold the common good and to protect and maintain the rights and best interests of its citizens, based on the ethical principles of beneficence, non-maleficence and justice.[844]

Biometrics: Upholding the Common Good?

While recognising that the use of biometric technologies raises particular ethical concerns regarding the rights and interests ordinarily held by all individuals, the possibility still arises that some of these rights may legitimately be limited or overridden where a given biometric application is deemed to be necessary to uphold some common good. Biometric applications are being implemented increasingly by government agencies under the pretext of upholding the common good as represented by policies of national and international security, public safety and law enforcement. Governments argue that allowing an individual to opt out of a national biometric programme could impact on the ability of the state to fulfil its responsibility to protect the rights of other citizens. Therefore, while the mandatory enrolment in specific biometric programmes may result in a limiting of a particular individual's right to privacy and autonomy in controlling the use and availability of his/her personal information and the right to opt out, the envisaged improvement in security and safety for everyone could be considered to justify the negative impact at the level of the individual.[845] In fact, Protocol No. 4 to the *Convention for the Protection of Human Rights and Fundamental Freedoms* states that "no restrictions shall be placed on the exercise of these rights [in relation to freedom of movement] other than such as are in accordance with law and are necessary in a democratic society in the interests of national security or public safety, for the maintenance of *ordre public*, for the prevention of crime, for the protection of health or morals, or for the protection of the rights and freedoms of others".[846] Protecting the security of the state and its citizens could be considered as sufficient reason to override certain individual rights, such as privacy and

---

839 Mill (1863) *op. cit.*

840 *ibid.*

841 Irish Council for Bioethics (2007) *op. cit.*

842 Donnolly M (2002). *Consent: Bridging the Gap between Doctor and Patient.* Cork University Press, Cork, 96p

843 Ryan FW (2002). *Constitutional Law.* Round Hall Ltd., Dublin 165p.

844 Irish Council for Bioethics (2007) *op. cit.*

845 Commission de l'éthique de la science et de la technologie (2008) *op. cit.*

846 Council of Europe 1963. Protocol No. 4 to the Convention for the Protection of Human Rights and Fundamental Freedoms securing certain rights and freedoms other than those already included in the Convention and in the first Protocol thereto. Strasbourg, 16.IX.1963.

autonomy in the short term in order to defend these and other rights overall in the longer term.[847] However, concerns have been raised that efforts to improve national security and fight the "war on terror" have been used to justify the increased securitisation of our everyday lives.[848] Whereas each state has a duty to protect its citizens from threats such as terrorism, the International Commission of Jurists (ICJ) have argued that any counterterrorism measures that are instigated should not undermine basic rights and international law.[849] For example, arguments in the interests of the common good and national security have been proffered previously as justification of actions that have violated human rights.[850,851] Furthermore, the ICJ did not agree that current claims of "exceptional" risks to safety and security required the implementation of exceptional countermeasures, which bypassed existing legislative frameworks.[852] The Council is concerned that the idea of upholding the common good may be over utilised as a means of justifying the implementation of numerous biometric and other applications, which can seriously impact on citizens' rights and civil liberties.

While an individual's rights and civil liberties are deserving of respect and are subject to legal protection, the Council recognises that these rights may be overridden by the state under certain circumstances for the benefit of the common good. However, the Council expresses concern that the argument of upholding the common good may be employed too readily as the reason for implementing particular programmes and applications. Therefore, given the limitations such programmes can place on an individual's civil liberties, there needs to be a proportionate justification and rationale for invoking the common good argument.

Proportionality

Justification and Necessity

An essential consideration when implementing a biometric application is determining whether or not the application is necessary and can be justified. There is a requirement to weigh the importance of society's need for the particular application, for example, to combat terrorism or identity theft, against concerns for individual rights and civil liberties.[853,854] Conflicts between individual rights and societal interests in relation to biometric applications may be resolved through the principle of proportionality. This principle requires that a balance be struck between the end a given application is hoping to achieve and the means by which this end is to be realised. In its most basic form the principle of proportionality, when relating to the implementation of biometric technologies states that any application or system "should balance its utility with the rights to privacy (personal, informational, *etc.*) of the involved

847 Clarke (1984) *op. cit.*
848 Muller BJ (2004). (Dis)Qualified Bodies: Securitization, Citizenship and "Identity Management". *Citizenship Studies* 8(3): 279–294.
849 International Commission of Jurists (2009) *op. cit.*
850 *ibid.*
851 Clarke (1984) *op. cit.*
852 International Commission of Jurists (2009) *op. cit.*
853 Woodward *et al.* (2001) *op. cit.*
854 Nuffield Council on Bioethics (2007) *op. cit.*

individuals".[855] It is therefore closely linked to the concept of data protection. According to the Organisation for Economic Cooperation and Development (OECD) "personal data should be relevant to the purposes for which they are to be used, and, to the extent necessary for those purposes, should be accurate, complete and kept up-to-date".[856] Proportionality, however, differs from data protection in at least one important aspect in that it establishes a balance between the usefulness of a specific application and its effects on privacy, whereas data protection imposes conditions on the collection and use of data.

Applying this principle requires a detailed assessment of the intended application, including the possible impact its implementation will have on the users of the system. For instance, biometric technologies have implications for privacy rights. According to the Ontario Information and Privacy Commissioner, biometrics requires people to "relinquish control over something that is highly personal and virtually immutable" and argues that caution is advisable.[857] Employing the principle of proportionality may not necessarily satisfy all competing interests; however, it encourages detailed examination and scrutiny of the proposed application. By its very nature, the principle requires that any decision is shown to be based on credible reasoning in order to be justified. Therefore, a justified intrusion on an individual's privacy needs to strike a reasonable balance between rights and utility. Accordingly, where the introduction of such large-scale biometric applications can be shown to represent a proportionate response to the particular problem at hand, then ethical concerns may be less likely to arise.

While certain biometrics programmes and applications may be justified at a national and international level, for example, immigration programmes, the use of biometrics may not always be considered proportional under other circumstances. Biometric technologies have been hailed as one of the many tools to combat serious anti-social behaviours, but they are increasingly being used in what is regarded as less "essential" situations. For instance, there are a number of cases in Ireland and abroad where institutions, such as schools and businesses, have installed biometric systems to record the time and attendance of pupils and staff. Furthermore, there are cases of recreational facilities, for example, gyms, golf clubs, casinos, theme parks and hotels, implementing biometric systems to verify the identity of their members and guests in a convenient manner. While the use of biometric technologies by organisations, such as nursing homes to protect the safety of vulnerable patients, for example those suffering from Alzheimer's disease, might be seen as a proportionate response to the risk of someone getting lost or injured, their use to track the movement of golf club members around a course might not. As one commentator has noted, "biometric identification

855 Iachello G and Abowd GD (2005). Privacy and Proportionality: Adapting Legal Evaluation Techniques to Inform Design in Ubiquitous Computing. *CHI* 2005, 2–7 April 2005, Portland Oregon, USA, p.91–100. Available online at: http://luci.ics.uci.edu/predeployment/websiteContent/weAreLuci/biographies/faculty/djp3/LocalCopy/p91-iachello.pdf, accessed on 21 November 2008.

856 Organisation for Economic Cooperation and Development (1980) *Guidelines on the Protection of Privacy and Transborder Flows of Personal Data*. OECD, Paris. Available online at: http://www.oecd.org/document/18/0,3343,en_2649_34255_1815186_1_1_1_1,00.html, accessed 23 October 2008.

857 Cavoukian (1999) *op. cit.*

procedures should be limited to those that need it, not promoted as a general panacea for security problems".[858]

In a discussion about the forensic use of bioinformation the Nuffield Council on Bioethics stated that if the desired objective can be achieved through a number of different means, then the least harmful of these means should be employed, *i.e.* the one that results in the least harm to the individual or society.[859] In the case of some biometric applications, different means could include using a different biometric modality, using a system based on localised rather than centralised storage of the biometric information, or even using a non-biometric alternative system altogether. Such a decision would take account of the suitability and appropriateness of the proposed means to meet the desired objective.

Assessing Proportionality

The European Commission's Article 29 Data Protection Working Party has noted that, according to EU data protection legislation, biometric data may be used only if it is adequate, relevant and not excessive.[860] In order to determine whether a specific application is proportionate, it must be shown that the objective of using it – for example, to uphold national security – justifies the burden it might place on personal rights.[861] Once the legitimacy of an application has been determined, its appropriateness, for example cost and practicability, as well as the appropriateness of alternative applications, must be evaluated.[862] Finally, the relevance of the application must be measured, *i.e.* the adequacy of the application to achieve a particular objective and its acceptability to all stakeholders must be ascertained.[863]

It is important to assess all the alternatives available, including the non-biometric options, the associated costs and implementation requirements, as well as the legal and ethical implications, before deciding whether or not to implement a biometric system. In addition, a number of additional factors also need to be taken into account in this regard, specifically:[864]

Environment: does the nature of the workplace require a high degree of security, for example, because of the value of the goods made or kept there or the sensitive information stored there?

Purpose: is a biometric system required to achieve the intended purpose or could a less intrusive method be used?

858 Crowley MG (2006). *Cyber crime and biometric authentication – the problem of privacy versus the protection of business assets.* 8p. Available online at: http://igneous.scis.ecu.edu.au/proceedings/2006/aism/Crowley%20-%20Cyber%20crime%20and%20biometric%20authentication%20the%20problem%20of%20privacy%20versus%20protection%20of%20business%20assets.pdf, accessed 21 November 2008.

859 Nuffield Council on Bioethics (2007) *op. cit.*

860 Article 29 Data Protection Working Party (2003) *op. cit.*

861 Iachello and Abowd (2005) *op. cit.*

862 *ibid.*

863 *ibid.*

864 Data Protection Commissioner (2008c). *Biometrics in the workplace.* Available online at: http://www.dataprotection.ie/docs/Biometrics_in_the_workplace./244.htm, accessed 13 May 2008.

Efficiency: is the introduction of a biometric system required to meet particular administrative requirements, which alternative, less intrusive methods have been unable to achieve?

Reliability: is a biometric system required to overcome a continuing problem of staff members impersonating each other, which other methods have failed to solve, for example, where people may be logging on to a network using someone else's password or PIN details to examine files they are not authorised to access.

For example, with a biometric application that only requires user verification (*i.e.* authentication of a claimed identity) such as a time and attendance system, it may be disproportionate to store the biometric information in a database when a smart card or match-on-card system could be used instead.[865] In such cases, when clocking in or out of work, each employee would scan his/her biometric identifier, which would then be compared with the template stored on the card. If the smart card also contained the individual's employee number this, as opposed to the biometric information, could be stored in a database with that individual's work time and attendance record. As has been discussed, verification-based biometric systems, which do not store the biometric information, generate less privacy concerns and allow the individual user to retain more control over his/her biometric information. Such systems are, thus, more likely to be considered proportionate and acceptable. Consequently, greater justification is usually required when implementing a biometric system that involves a centralised database, but such systems are often considered proportionate for issues such as national and international border management and security, benefit and entitlement fraud and law enforcement.

Finally, having decided to store biometric and/or associated personal information in a database, an appropriate data retention policy is required, commensurate with the importance of the application in question.[866,867,868] For example, in the majority of cases, it is considered reasonable to retain the information while the individual is still using the biometric system in question, for example, if he/she is still employed by that company or attending that school. However, if that individual is no longer employed by that company it could be considered inappropriate to retain his/her biometric information on the database.[869,870] Nonetheless, large-scale, government-based applications may require longer retention policies, but again such policies require justification. In the case of the US-VISIT programme the information retention period is 75 years, which is seen as the "minimal period necessary to carry out DHS national security, law enforcement, immigration, intelligence and other mission-related functions."[871]

---

865 As noted above, a match-on-card system is one where the biometric information (*i.e.* raw image or template), as well as the feature extraction and matching modules are all stored and conducted on a smart card.

866 Data Protection Commissioner (2008c) *op. cit.*

867 Snijder (2007) *op. cit.*

868 Commission de l'éthique de la science et de la technologie (2008) *op. cit.*

869 International Biometric Group (2007a) *op. cit.*

870 Data Protection Commissioner (2008c) *op. cit.*

871 Department of Homeland Security (2007) *op. cit.*

In the Council's opinion, the justification of implementing a biometric application is reliant on the application being considered proportionate. Biometric applications should therefore be assessed on a case-by-case basis, which involves a consideration of the relevance and necessity of employing biometric technologies, given the proposed purpose of the system, the environment in which it will be used, and the level of efficiency and degree of reliability required to achieve the proposed purpose.

Trust and Transparency

In order to minimise the ethical concerns raised by biometric technologies, it has been argued that the operation and management of biometric applications, whether compulsory or voluntary, needs to be transparent and conducted in accordance with the appropriate regulations and with respect for fundamental ethical principles, human rights and civil liberties.[872,873,874] The issues of trust and transparency play an important role in the arenas of national security, law enforcement and border integrity generally and specifically in their utilisation of biometric technologies. Using physiological characteristics to identify oneself is a relatively new concept for people to understand and accept as normal practice. In order for biometric technologies to be used and accepted, people must firstly trust the systems themselves and secondly those who operate them. Public trust would need to be at such a level that people would be willing to replace traditional identification methods, for example PINs and passwords with iris scans and fingerprints. It is, therefore, imperative for organisations wishing to roll out biometric identification systems to operate in a transparent fashion, and to educate and inform people about the technology itself as well as the implications that might arise from its use.

In recent years, a significant amount of research into the acceptance and trust of biometric technologies and those who operate them for identification purposes has been undertaken. A number of studies reported considerable apprehension in relation to privacy,[875] with this concern being echoed in both mainstream media and specialised publications. Alleviating the public's anxiety regarding their right to privacy is a challenge for governments and commercial organisations seeking to introduce large-scale biometric systems because the more concerned people are about the implications for privacy the less likely they are to accept the technology. In a 2006 study, which investigated the factors affecting the adoption of biometrics, 71 per cent of respondents identified privacy as a significant concern.[876] The purpose of another study undertaken in 2006 was to assess the attitudes of people in the Americas, Europe and Asia-Pacific to various methods of "identity management". The research, which was undertaken on behalf of the Unisys Corporation, asserted that identity management would only work if there

---

872 van der Ploeg (2005a) *op. cit.*

873 Commission de l'éthique de la science et de la technologie (2008) *op. cit.*

874 Lodge (2007) *op. cit.*

875 Troitzsch H, Eschenburg F, Bente G, Krämer N, Lylykangas J, Vuorinen K and Surakka V (2005). *Deliverable D6.5. Introduction of a Multi-Modular Acceptance and Usability Questionnaire.* Version 1.2, 92p.

876 Dike-Anyiam B and Rehmani Q (2006). Biometric vs. Password Authentication: A User's Perspective. *The Journal of Information Warfare* 5(1): 33–45.

was complete public cooperation and acceptance of the technology used and that resistance would occur if the technology were seen to encroach on privacy rights.[877] However, privacy was not the only factor that caused concern. When asked why they would not consider using biometrics, 74 per cent of respondents cited suspicion of how the technology works and 32 per cent said they were fearful of their information being accessed/abused by third parties. In an open section at the end of the survey, where people could express further opinions, many people said that, in order for identity management to be accepted, those implementing it would need to have "impeccable" standards of ethics and transparency.[878]

Focus groups conducted on behalf of the Council by Red C in March 2008 (see Appendix A) demonstrated that there was a general acceptance and understanding of the need for heightened security measures in the interest of national and personal security, as well as a perception that biometric identification would add an extra layer of protection to information. However, there was also an acknowledgement that breaches of security were always a possibility. Awareness was based on the exposure respondents had had to TV (television) programmes and films about biometrics, as well as their own direct experience of systems, for example, at work or while travelling. However, there was a low level of understanding regarding how data is generated from biometric characteristics and how it is used. In terms of data transfer, there was a broad acceptance of the transmission of personal data between government agencies so long as the data was relevant to the needs of that agency. For instance, while it would be acceptable for a hospital to be given health insurance details, it would be unacceptable for a library to know a person's blood group. However, it should be noted that the focus groups were made up of a very small number of people who participated voluntarily, and it cannot be assumed that their views and perceptions are representative of the wider public.

One challenge facing parties wishing to implement biometric identification systems is the inability of legislative and ethical controls to keep pace with rapidly developing technologies, which citizens might view as expensive, overly intrusive and unnecessary. Another issue, which has implications for the introduction of biometric systems, is often referred to as a "decline in trust". One body of evidence suggests that, over the last number of years, society has become less trusting in general and of government in particular.[879,880,881] Current standards regarding security highlight the paradox in the modern democratic system. In order to be deemed democratic a state is supposed to be guided by the will of the people; however, many biometric systems introduced by national governments are compulsory and were implemented without consultation with citizens. Therefore, if public trust is to be gained, it might be appropriate for governments to show accountability and be transparent regarding their use of biometric technologies. In political terms the concepts of transparency and accountability are important for ensuring good governance. While accountability enables feedback only after a decision has been made or action taken, transparency allows for feedback during the process

877 Ponemon (2006) *op. cit.*

878 *ibid.*

879 Perri 6, Lasky K and Fletcher A (1998). *The Future of Privacy. Volume 2, Public trust and the use of private information.* Demos, London, 144p.

880 O'Hara K (2004). *Trust: From Socrates to Spin*, Icon Books, Cambridge 256p.

881 O'Neill O (2002). *A Question of Trust: The BBC Reith Lectures*, Cambridge University Press, Cambridge, 108p.

of making decisions and taking actions. Biometric systems, which are already becoming part of Irish life, have often been introduced without consultation. While, there is a general acceptance of these new systems, there is a potential for serious public backlash if major errors were to occur, for example, loss of data or unauthorised access to a database. Transparency might be displayed in a number of ways, including providing education and information, supporting strict data controls, establishing a formal procedure for audit and providing sufficient mechanisms for complaints and redress. As one commentator notes, transparency tools define the principles by which states and private organisations should behave in relation to their constituents, stating that "transparency tools tend to make the powerful transparent and accountable: they allow us 'to watch the watchdogs'".[882]

Transparency requires open debate between all parties who will be involved in the system to help clarify any issues and concerns that may arise. It is considered essential that such debate occurs prior to the establishment of the proposed biometric programme, particularly given the ethical concerns that have been raised. For example, if the users of the system consider the level of invasiveness of a biometric technology disproportionate, this could discourage acceptance of the system and could damage the level of trust between the parties. In addition, the appropriate regulations and safeguards also need to be outlined in advance to make sure both system users and operators are aware of their rights and responsibilities. Therefore, should any issues arise, those affected will have the opportunity to rectify the situation.

On 17 May 2009, Switzerland held a referendum on a proposed new law on biometric passports. Voters (50.14 per cent) supported the implementation of biometric passports, which will contain an individual's photograph plus two fingerprint images.[883] Low voter turnout (38 per cent) coupled with the closeness of the result was seen by some sections of the media as a sign of indecisiveness or scepticism among the population.[884] The referendum was called for after a broad political coalition challenged the parliament's decision to introduce biometric passports and to develop a central registry of fingerprints.[885] Opponents of biometric passports raised concerns around individual privacy, security and data protection, particularly in relation to the central fingerprint registry.[886,887] Many of these concerns centred on the potential risk of abuse posed by hackers and also the possibility of the police accessing the fingerprint registry as part of their investigations.[888] While welcoming the vote in favour of biometric passports, the Swiss Justice Minister Eveline Widmer-Schlumpf promised to take the privacy concerns

882 Gutwirth S (2007). Biometrics between opacity and transparency. *Annali dell Institute Superiore di Sanitá* 43(1): 61–65.

883 European Digital Rights (2009). *Lucky Win For The Swiss Biometric Passports.* Available online at: http://www.edri.org/edri-gram/number7.20/swiss-biometric-passports, accessed 1 July 2009.

884 *ibid.*

885 Geiser U (2009a). *Swiss vote on introduction of biometric passports.* Available online at: http://www.swissinfo.ch/eng/front/Swiss_vote_on_introduction_of_biometric_passports.html?siteSect=105&sid=10707459&rss=true&ty=st, accessed 1 July 2009.

886 Geiser U (2009b). Passport Vote Wins Majority and Puzzles Experts. *The Journal of Turkish Weekly* 17 May 2009. Available online at: http://www.turkishweekly.net/news/77082/-passport-vote-wins-majority-and-puzzles-experts.html, accessed 1 July 2009.

887 Geiser U (2009a) *op. cit.*

888 *ibid.*

raised into consideration and to ensure the security of personal data.[889] Minister Widmer-Schlumpf stated that current laws would prevent the police from accessing the fingerprint registry and any attempts to change such laws would require the approval of parliament.[890]

As biometric identification technology becomes more prevalent in day-to-day life, the issues of potential data misuse and high-tech surveillance become paramount. People often have misconceptions about the capabilities of new technologies, not least biometrics. Kush Wadhwa has stated that there is already a huge amount of misinformation in circulation with regard to biometrics.[891] For example, Wadhwa believes it is very important to highlight the difference between how traditional identification methods work and how biometric systems work. Unlike traditional identification, for example, using a password, which is either right or wrong, biometric systems are probabilistic in nature,[892] with the correct answer being based on the similarity of the compared templates or images in relation to some operator defined threshold. Failure to explain the inner workings of systems in an attempt to make them easier to understand, or indeed why the systems are being implemented, will do little to dispel myths and fears. Perhaps a more valuable approach may be to ensure people understand how these systems work, including how the biometric features are captured, extracted, stored and used. Providing adequate education and ensuring appropriate policy making would diminish concerns and misunderstandings regarding the implementation of biometric technologies.

The Council believes that increased transparency and honesty regarding biometric technologies, applications, the use to which an individual's biometric information will be put and who will have access to this information is essential in garnering the trust and acceptance of the intended users of these systems. This includes providing information on the most up to date independent research and developments in biometrics and accurate information on the role the biometric application will play in resolving the particular problem at hand. An important aspect of this transparency is the need for a full and frank debate on the issues raised by all parties who will be involved in the proposed application, prior to the establishment of the proposed programme. This is considered particularly important for applications where participation will be mandatory.

---

889 Geiser U (2009b) *op. cit.*

890 *ibid.*

891 Biometric Information Technology Ethics (2005) *op. cit.*

892 *ibid.*

CHAPTER 4
BIOMETRICS LEGISLATION/ REGULATION

Chapter 4: Biometrics Legislation/Regulation

Biometric information has a unique quality because of its connection to physiological and behavioural characteristics, which allows for the identification of an individual. The rapid development and increased application of biometric technologies in recent years has raised concerns with respect to the protection of citizens' fundamental rights and freedoms, in particular their right to privacy. Privacy refers to our right to control access to ourselves and to our personal information. It is a fundamental right recognised in many international instruments and regulations, including the UN *Universal Declaration of Human Rights,*[893] the Council of Europe's *Convention for the Protection of Individuals with Regard to Automatic Processing of Personal Data*[894] and the *International Covenant on Civil and Political Rights.*[895]

Indeed, there has been a long history of legislative protection for an individual's right to privacy. Privacy laws can be traced back to 1361, when the *Justices of the Peace Act* in England provided for the arrest of peeping Toms and eavesdroppers.[896] During the succeeding centuries a number of countries enacted privacy legislation. However, it was the introduction of the UN *Universal Declaration of Human Rights*, which provided the yardstick for modern privacy laws. According to Article 12: "No one shall be subjected to arbitrary interference with his privacy, family, home or correspondence, nor to attacks upon his honour and reputation. Everyone has the right to the protection of the law against such interference or attacks."[897] This was followed soon after by the *Convention for the Protection of Human Rights and Fundamental Freedoms* in 1950, Article 8(1) of which states that "Everyone has the right to respect for his private and family life, his home and his correspondence."[898] From this Convention, the European Commission of Human Rights and the European Court of Human Rights (ECHR) were created to oversee the protection of personal privacy rights. By virtue of the *European Convention on Human Rights Act 2003,*[899] the Convention is now part of Irish law.

The right to privacy is thus recognised worldwide as being integral to a fair and stable society. Indeed, almost every country in the world has either made provisions for the protection of privacy in their constitutions or has enacted specific legislation. It should be noted, however, that the right to privacy is not absolute and has been overridden in the interest of the common good or "public interest". The most common restrictions placed on privacy for the benefit of public interest include the prevention of harm to others, the prevention/detection/prosecution

893 United Nations, *Universal Declaration of Human Rights* (1948). Available online at: http://www.unhchr.ch/udhr/lang/eng.htm, accessed 20 October 2008.

894 Council of Europe (1981). *Europe Convention for the Protection of Individuals with Regards to Automatic Processing of Personal Data,* 28.I.1981. Available online at http://conventions.coe.int/Treaty/en/Treaties/Html/108.htm, accessed 30 January 2008.

895 United Nations (1976). *International Covenant on Civil and Political Rights.* Available online at: http://www.unhchr.ch/html/menu3/b/a_ccpr.htm, accessed 5 March 2008.

896 Henderson SC and Snyder CA (1999). Personal information privacy: implications for MIS managers. *Information & Management* 36(4): 213–220.

897 United Nations (1948) *op. cit.*

898 Council of Europe (1950–1998). *Convention for the Protection of Human Rights and Fundamental Freedoms as amended by Protocol No. 11.*

899 European Convention on Human Rights Act (2003:20).

of crime, where a litigant loses or waives his/her privacy rights and for purposes where statute requires disclosure.[900]

Article 8(2) of the *European Convention for the Protection of Human Rights and Fundamental Freedoms* states that:

> There shall be no interference by a public authority with the exercise of this right except such as in accordance with the law and is necessary in a democratic society in the interests of national security, public safety or the economic well-being of the country, the prevention of disorder or crime, for the protection of health or morals, or for the protection of the rights and freedoms of others.[901]

Similarly, the *Council of Europe Convention for the Protection of Human Rights and Dignity of the Human Being with Regard to the Application of Biology and Medicine: Convention on Human Rights and Biomedicine* (Oviedo)[902] places restrictions on privacy rights. Article 26 states that:

> No restrictions shall be placed on the exercise of the rights and protective provisions contained in this convention other than such as are prescribed by law and are necessary in a democratic society in the interest of public safety, for the prevention of crime, for the protection of public health or for the protection of the rights and freedoms of others.

It should be noted that Ireland has not signed the *Convention on Human Rights and Biomedicine*. In addition to the aforementioned conventions, a body of jurisprudence has also been developed.

The ECHR recently found against the UK in *S. and Marper* v. *The United Kingdom*.[903] The applicants complained under Articles 8 and 14 of the *Convention for the Protection of Human Rights and Fundamental Freedoms* that the UK police service had continued to retain their fingerprints, cellular samples and DNA profiles after the criminal proceedings against them had ended with an acquittal (Mr S.) and discontinued (Mr Marper). The case was brought to the ECHR in 2008,[904] where the previous rulings in the UK were overturned and the ECHR concluded that there had been a violation of Article 8 of the *Convention*. In its judgment the ECHR ruled that:

> ... the blanket and indiscriminate nature of the powers of retention of the fingerprints, cellular samples and DNA profiles of persons suspected but not convicted of offences, as applied in the case of the present applicants, fails to

900 Sheikh AA (2008). *The Data Protection Acts 1988 and 2003: Some Implications for Public Health and Medical Research*, Health Research Board, Dublin, 130p.

901 Council of Europe (1950-1998) *op. cit.*

902 Council of Europe (1997). *Convention for the Protection of Human Rights and Dignity of the Human Being with Regard to the Application of Biology and Medicine: Convention on Human Rights and Biomedicine*, Oviedo.

903 *S. and Marper* v. *United Kingdom* [2008] ECHR 30562/04 [Grand Chamber].

904 In 2002 the UK Administrative Court rejected both applications: a decision, which was upheld by the Court of Appeal in September of that year. In July 2004 S. and Marper's appeal was dismissed by the House of Lords.

> strike a fair balance between the competing public and private interests and that the respondent State has overstepped any acceptable margin of appreciation in this regard. Accordingly, the retention at issue constitutes a disproportionate interference with the applicants' right to respect for private life and cannot be regarded as necessary in a democratic society.

In an Irish context there are two cases in particular which highlight the need to balance the right to privacy with the exigencies of the common good.[905] In *Kennedy and Arnold* v. *Ireland* (where the Supreme Court ruled that illegal wiretapping was in breach of the Constitution), Hamilton P. stated that the right to privacy is "not an unqualified right" and "its exercise may be restricted by the constitutional rights of others, by the requirements of the common good and is subject to the requirements of public order and morality."[906] In *Haughey* v. *Moriarty*[907] the Supreme Court concluded that the establishment of a tribunal of inquiry was justified because, while it encroached the plaintiff's privacy right it did so in order to facilitate investigations into matters regarded as being of the utmost public importance and in the interest of the common good. According to Hamilton CJ:

> The exigencies of the common good require that matters considered by both Houses of the Oireachtas to be of urgent public importance be inquired into, particularly when such inquires are necessary to preserve the purity and integrity of our public life without which a successful democracy is impossible.[908]

The right to privacy may also be limited by statute. *The Criminal Justice (Surveillance) Act 2009* states that its main purpose lies in the "prevention and detection of serious crime and in safeguarding the security of the State against subversion and terrorism".[909] Effectively, this Act allows the Garda Síochána,[910] the Defence Forces and the Revenue Commissioners to use surveillance devices, for example, video cameras, audio bugs or tracking equipment to monitor, observe, listen or make a recording "of a particular person or group of persons or their movements, activities or communications". The legislation allows evidence gathered using these surveillance techniques to be used in criminal prosecutions for serious offences, which are punishable by prison terms of at least five years.

While privacy legislation is well ensconced in most jurisdictions, there is relatively little legislation in Europe or globally dealing specifically with biometric technologies. In recent years, biometrics have become ubiquitous in society, in both the public and private spheres. This escalation has necessitated the introduction of a new legal framework, which is tailored to deal specifically with biometrics and its associated issues, as well as the updating of

---

905 Delany H (2008). *The Right to Privacy. A Doctrinal and Comparative Analysis.* Thomson Round Hall, Dublin, 352p.

906 *Kennedy and Arnold* v. *Ireland* [1987] I.R. 587.

907 *Haughey* v. *Moriarty* [1999] 3 I.R. 1.

908 *ibid.*

909 *The Criminal Justice (Surveillance) Act 2009*

910 The Garda Síochána are the Irish police force.

existing legislation, both nationally and internationally. It remains unclear whether current data protection laws encompass biometric data; therefore, a number of countries have begun crafting new legislation in this area.

Europe

As mentioned previously, there is a dearth of legislation in Europe, which specifically relates to biometric technologies. In 2004, the Council of Europe passed Regulation EC 2252/2004[911] for the purposes of harmonising security standards for passports and ensuring a reliable link between individuals and their travel documents by integrating biometric identifiers (facial image and fingerprints) into passports. The regulation also ensures that Member States meet the requirements of the US visa waiver programme (US-VISIT). Article 4 of the regulation states that the purpose of biometric passports is to verify the authenticity of the document and the identity of the holder "by means of directly available comparable features". According to Article 6, Member States must have introduced biometric facial images into new passports within 18 months after the enactment of the regulation. For fingerprint enabled passports the timeframe was initially set for 36 months after enactment, with a deadline for February 2008.

In May 2009, the European Council introduced Regulation (EC) 444/2009 in order to amend Regulation 2252/2004.[912] Regulation 444/2009 states that two particular groups shall be exempt from having to provide fingerprints, *i.e.* children under the age of 12 years and persons where fingerprinting is physically impossible. The regulation also provides that where fingerprinting the designated fingers is temporarily impossible, Member States shall allow the fingerprinting of other fingers. Where it is temporarily impossible to fingerprint any of the fingers, the person may be issued with a temporary passport, which would be valid for up to 12 months.

Regulation 444/2009 provides that biometrics may be taken only by "qualified and duly authorised staff" of the national authorities responsible for issuing passports and travel documents and that they be collected in accordance with the Council of Europe's *Convention for the Protection of Human Rights and Fundamental Freedoms* and the *United Nations Convention on the Rights of the Child*, hence, "guaranteeing the dignity of the person concerned".[913] Furthermore, the regulation calls for a large-scale and in-depth study, which will examine the reliability and technical feasibility of using the fingerprints of children under the age of 12 years, including a comparison of false rejection rates in each Member State. A report based on this research must be submitted to the European Parliament and Council no later than June 2012.

---

911 Council Regulation (EC) No. 2252/2004 of 13 December 2004 *on standards for security features and biometrics in passports and travel documents issued by member states.*

912 Regulation (EC) No. 444/2009 of the European Parliament and of the Council of 28 May 2009 amending Council Regulation (EC) No. 2252/2004 on standards for security features and biometrics in passports and travel documents issued by Member States.

913 *ibid.*

It should be noted that while the majority of EU Member States are legally bound by Regulation 2252/2004 and the later Regulation 444/2009, Ireland[914] and the UK[915] are not. This is because the Regulations represent a development of provisions of the Schengen *acquis*, in which neither country participated. Despite abstaining from the Schengen agreement, new passports issued by Ireland and the UK do comply with European and US-VISIT requirements.

In 1970, the world's first data protection law was passed in Hesse, Germany.[916] Other national laws soon followed (*e.g.* Sweden,[917] the US,[918] Germany,[919] Austria,[920] Denmark[921] and France[922]). During the 1980s, two international instruments calling for the protection of personal data from the point of collection to storage and disclosure were adopted. The first was the Council of Europe's *Convention for the Protection of Individuals with Regard to Automatic Processing of Personal Data.*[923] The aim of the Convention was to strengthen data protection in the light of the increased use of computers for administrative purposes and the rise in the transborder flow of automated personal information.[924] The second was a set of recommendations by the OECD entitled *Guidelines Governing the Protection of Privacy & Transborder Flows of Personal Data.*[925] The key principles embodied in the guidelines are that personal information must be:

- collected fairly and lawfully;
- used only for the purpose specified during collection;
- adequate, relevant and not excessive to that purpose;
- accurate and up to date;
- accessible in order for individuals to verify or correct their data;
- stored securely;
- disposed of once the specified purpose has been achieved.

In 1995, the EU adopted the Data Protection Directive,[926] which was designed to harmonise standards for the fair processing of personal data and ensure its free movement between Member States. The Directive protects a number of rights, for example, the right to rectify inaccurate data, the right of recourse in the event of illegal processing and the right to refuse

---

914 Council Decision of 28 February 2002, concerning Ireland's request to take part in some of the provisions of the Schengen *acquis* (2002/192/EC).

915 Council Decision of 29 May 2000, concerning the request of the United Kingdom of Great Britain and Northern Ireland to take part in some of the provisions of the Schengen *acquis* (2000/365/EC).

916 Hessisches Datenschutzgesetz (The Hesse Data Protection Act), Gesetz und Verordungsblatt I (1970), 625.

917 The Data Act (1973).

918 Privacy Act (1974).

919 Federal Data Protection Act (1977).

920 Data Protection Act (1978).

921 Private Registers Act (1978) and Public Authorities' Registers Act (1978).

922 Act n°78–17 of 6 January 1978 on Data Processing, Data Files and Individual Liberties.

923 Council of Europe (1981) *op. cit.*

924 In 2001, an additional protocol was added to the Convention, which called for the establishment of supervisory authorities to ensure compliance in Member States and which provided for the transborder flow of information to recipients outside of the Convention's jurisdiction and where an adequate level of data protection is ensured. Council for Europe, Additional Protocol to the *Convention for the Protection of Individuals with regard to Automatic Processing of Personal Data regarding supervisory authorities and transborder flows*, Strasbourg, 8.X1.2001.

925 Organisation for Economic Cooperation and Development (1980) *op. cit.*

926 *Directive 95/46/EC of the European Parliament and of the Council of 24 October 1995 on the protection of individuals with regard to the processing of personal data and on the free movement of such data.*

consent for the use of data in certain circumstances. The 1995 Directive also requires Member States to ensure personal information relating to European citizens is afforded the same level of protection when it is transferred to non-Member States. Article 28 of the Directive called for the establishment of an independent supervisory authority (Data Protection Commissioner) to oversee data protection in each Member State. These supervisory bodies have a number of powers, including requesting the government to consult them when drawing up legislation relating to data processing, conducting investigations into alleged data protection breaches and initiating legal proceedings, *etc.* Under Article 29, a working party, comprised of representatives from the supervisory authority in each Member State, was established. The Article 29 Working Party (WP29) is an independent advisory body given a number of tasks, including examining questions regarding the application of national data protection measures, the preparation of opinions on the level of data protection within the EU and in third countries, advising the European Commission on new measures which might be adopted and making recommendations on all matters relating to data protection. WP29 has issued numerous opinions relating to issues, such as the introduction of EU wide data retention requirements, the transfer of travellers' personal information to US authorities and the introduction of biometrics into passports and visas. In its opinion on biometrics in passports and travel documents, WP29 called for a number of safeguards, including restricting the use of biometrics in passports and other travel documents for the purpose of verification by comparing the data in the document with the data provided by the holder, a guarantee that the passports, *etc.* of European citizens could not be read by unauthorised parties and that only competent authorities, appointed by Member States, should have access to biometric data.[927]

In April 2008 the Information Commissioner's Office (ICO) contracted the RAND Corporation to undertake a review of the 1995 Data Protection Directive. In its evaluation, RAND highlights the dramatic changes that have occurred in the way personal information is collected, processed and used. The report emphasises that the Directive was written during a period when data processing entailed the use of filing systems and computer mainframes and that its main objective was to harmonise existing regulations to safeguard informational privacy rights rather than to create a legal framework to cope with future data processing and privacy challenges.[928] The report concludes that the Directive, as it currently exists, with its roots in a "largely static and less globalised environment", will not suffice in the long term.[929] The RAND report sets out a number of recommendations for maximising the effectiveness of the current legislative arrangements as well as proposals for a regulatory framework, which would anticipate future technological developments.

In 2000, the European Parliament, Council and the Commission introduced the *Charter of Fundamental Rights of the European Union.*[930] Article 8 of the Charter states that everyone has

927 Article 29 Data Protection Working Party (2005). *Opinion on Implementing the Council Regulation (EC) No 2252/2004 of 13 December 2004 on standards for security features and biometrics in passports and travel documents issued by Member States.* European Commission, Brussels, 12p. Available online at: http://ec.europa.eu/justice_home/fsj/privacy/docs/wpdocs/2005/wp112_en.pdf, accessed 17 November 2008.

928 Robinson N, Graux H, Botterman M, Valeri L (2009). *Review of the European Data Protection Directive,* RAND Corporation, Cambridge, 82p.

929 *ibid.*

930 Official Journal of the European Communities (2000). *Charter of Fundamental Rights of the European Union* (2000/c 364/01).

the right to protection of their personal data and that "such data must be processed fairly for specified purposes and on the basis of the consent of the person concerned or some other legitimate basis laid down by law. Everyone has the right of access to data which has been collected concerning him or her, and the right to have it rectified."[931] Article 52 of the Charter clarifies the scope of guaranteed rights by stating that:

> Any limitation on the exercise of the rights and freedoms recognised by this Charter must be provided for by law and respect the essence of those rights and freedoms. Subject to the principle of proportionality, limitations may be made only if they are necessary and genuinely meet the objectives of general interest recognised by the Union or the need to protect the rights and freedoms of others.[932]

In 2001 the European Data Protection Supervisor (EDPS) was established under Article 41 of Regulation (EC) 45/2001.[933] The EDPS's mission is to ensure that EU institutions and bodies respect the fundamental right to protection of personal data. The obligations listed in this regulation are similar to those set out in the 1995 Data Protection Directive. The EDPS has published a number of opinions on such topics as security features and biometrics in passports, European passenger name records and RFID, and has reinforced a number of opinions expressed by the WP29. For instance, in its opinion on a proposal by the EU Commission to amend Regulation 2252/2204[934] on standards for security features in passports and travel documents, the EDPS made a number of recommendations, such as: the age limit for proposed exemptions for children and the elderly should be altered to match limits already adopted for Eurodac and the US-VISIT systems, *i.e.* from 6 years to 14 years for children and to 79 years for the elderly; that those exempted should not experience any stigmatisation or discrimination as a result; that measures to harmonise the types of documents required for the issuing of passports should be introduced among Member States, in order to enhance data security; and harmonisation measures should be introduced to implement the decentralised storage of biometric data for Member States' passports.[935]

In May 2005, the European Commission launched a five-year Action Plan for Freedom, Justice and Security, which included detailed proposals for action on terrorism, migration management, visa policies, asylum, privacy and security, and organised crime and criminal justice. The Action Plan takes the overall priorities for Freedom, Justice and Security, which were laid out in the Hague Programme that was endorsed by the European Council in November 2004. According to Section 1.7.2 of the Hague Programme:

---

931 *ibid.*

932 *ibid.*

933 Regulation (EC) No 45/2001 of the European Parliament and the Council of 18 December 2000 *On the protection of individuals with regard to the processing of personal data by the Community institutions and bodies and on the free movement of such data.*

934 28/06/2006. C(2006) 2909. Commission Decision establishing the technical specifications on the standards for security features and biometrics in passports and travel documents issued by Member States. Available online at: http://ec.europa.eu/justice_home/doc_centre/freetravel/documents/doc/c_2006_2909_prov_en.pdf, accessed 17 November 2008.

935 European Data Protection Supervisor (2008). Opinion of the European Data Protection Supervisor on the proposal for a Regulation of the European Parliament and of the Council amending Council Regulation (EC) No 2252/2004 on standards for security features and biometrics in passports and travel documents issued by Member States, (2008/C 200/01). *Official Journal of the European Union* C200: 1-5. Available online at: http://www.edps.europa.eu/EDPSWEB/webdav/site/mySite/shared/Documents/Consultation/Opinions/2008/08-03-26_Biometrics_passports_EN.pdf, accessed 17 November 2008.

> The management of migration flows, including the fight against illegal immigration should be strengthened by establishing a continuum of security measures that effectively links visa application procedures and entry and exit procedures at external border crossings. Such measures are also of importance for the prevention and control of crime, in particular terrorism. In order to achieve this, a coherent approach and harmonised solutions in the EU on biometric identifiers and data are necessary.[936]

The Hague Programme also calls for an investigation into how to maximise the effectiveness and interoperability of EU information systems, *i.e.* the second generation Schengen Information System (SIS II), the Visa Information System (VIS) and Eurodac, in tackling illegal immigration and improving border controls, taking into account the need to strike the right balance between law-enforcement purposes and safeguarding the fundamental rights of individuals.[937]

Ireland

Ireland is subject to most EU directives and treaties, the exception in this case being Regulation 2252/2004, and has, therefore, enacted national data protection as well as human rights legislation. While not mentioned specifically, it is held that an individual's right to privacy is protected under Article 40.3.2 of the Irish Constitution, which proclaims that "The State shall, in particular, by its laws protect as best it may from unjust attack and, in the case of injustice done, vindicate the life, person, good name and property rights of every citizen".[938] As noted above, privacy was first accepted as a constitutional right in the case of *McGee* v. *Attorney General*,[939] which recognised the right to marital privacy and later in *Kennedy and Arnold* v. *Ireland*.[940]

In Ireland, a person's right to privacy is also provided for in the *Data Protection Acts, 1988*,[941] *2003*.[942] The Acts define personal data as "data relating to a living individual who is or who can be identified either from the data or from the data in conjunction with other information that is in, or is likely to come into, the possession of the data controller". According to this legislation, personal data:

- shall be obtained for one or more specified, explicit and legitimate purposes;
- shall not be further processed in a manner incompatible with the original purposes;
- shall be adequate, relevant and not excessive;

---

936 Council of the European Union (2004). *The Hague Programme: strengthening freedom, security and justice in the European Union*, 16054/04. Council of the European Union, Brussels.
Available online at: http://ec.europa.eu/justice_home/doc_centre/doc/hague_programme_en.pdf, accessed 17 November 2008.

937 *ibid.*

938 *Bunreacht na hÉireann, Constitution of Ireland*, 1937–2004, Article 40.3.2.

939 *McGee* v. *Attorney General* [1974] IR 284.

940 *Kennedy and Arnold* v. *Ireland* [1987] IR 587; 1988 ILRM 472.

941 The Data Protection Act (1988:25) was enacted in order to implement the Council of Europe *Convention for the Protection of Individuals with Regards to Automatic Processing of Personal Data* (1981).

942 Data Protection (Amendment) Act 2003 (Commencement) Order (2007:656) was enacted in order to comply with the European Data Protection Directive of 1995.

- shall not be kept longer than is necessary; and
- shall not be disclosed to any third party except in a manner compatible with the original purpose.

The Act also requires that data controllers install appropriate security measures against unauthorised access, alteration, disclosure or destruction of data.

While Irish data protection legislation does not specifically mention biometric data, the Office of the Data Protection Commissioner has reported an increased number of enquiries and complaints regarding the use of biometric technologies for a range of purposes.[943] In particular, the Commissioner expressed concern relating to the introduction of biometric technologies to record time and attendance in schools and workplaces – the chief of which was the possibility that people would become "desensitised" to biometrics and would, consequently, be less aware of their privacy rights. While the Office of the Data Protection Commissioner stated that it would review the use of biometric technologies in workplaces on a case-by-case basis,[944] it issued a set of guidelines for schools, laying out their responsibilities as data controllers. The key instructions of this document are that either students or (in the case of those under the age of 18) their parents/guardians give explicit consent before biometric data is taken and that students "be given a clear and unambiguous right to opt out of a biometric system without penalty".[945] Indeed, the Data Protection Commissioner states that "the processing of sensitive personal data through the use of a biometric system is not necessary to meet the requirements of the *Education (Welfare) Act*, 2000 in respect of recording student attendance."[946]

There are two pieces of Irish legislation that mention biometrics specifically, *i.e.* the *Passport Act 2008*[947] and the *Immigration, Residence and Protection Bill 2008*.[948] Section 8(1) of the *Passport Act* states that, subject to data protection legislation, biometric data may be processed in respect of issuing passports. Section 8(2) states that the Minister for Justice, Equality and Law Reform may select "such persons as he or she sees fit" to process biometric data. Section 13(1) of the Act states that, "having regard to international standards and practice regarding the issue and form of passports generally", biometric information may be entered into a passport, in order to help identify the passport holder. The Act does not stipulate what types of biometric information might be required.

Section 108(1) of the *Immigration, Residence and Protection Bill* provides that a foreign national must furnish the Minister for Justice, Equality and Law Reform, an immigration officer, a member of the Garda Síochána or a member of the civilian staff of the Garda Síochána with "such biometric information in such a manner as may reasonably be required for the purposes of the performance of his or her functions" under the Bill. The Bill defines biometric information as "information about the distinctive physical characteristics of a person". However, the Bill,

---

943 Data Protection Commissioner (2008a) *op. cit.*

944 Data Protection Commissioner (2008c) *op. cit.*

945 Data Protection Commissioner (2007) *op. cit.*

946 *ibid.*

947 *Passport Act* (2008).

948 *Immigration, Residence and Protection Bill* (2008).

like the *Passport Act* does not specify what types of biometric data might be required. Section 108(2) states that anyone who refuses to comply with the above-mentioned requirement will be guilty of an offence. These two provisions relate to any foreign nationals thought to be over the age of 14. Anyone regarded as being under the age of 14 will be requested to provide biometric information only in the presence of a parent/guardian. Section 108(7) allows the Minister to establish and maintain a record of biometric information for the purpose of storage and comparison (whether for the purposes of this Bill or for other laws). Section 107 of the Bill allows the Minister to "provide information about a foreign national to and receive such information from another state". At the time of writing, the *Immigration, Residence and Protection Bill* was still being discussed by the government's Select Committee on Justice, Equality, Defence and Women's Rights. The Irish Human Rights Commission has made a number of recommendations in relation to the Bill, including limiting the types of biometric data to be collected, defining more clearly the purposes for which biometric data can be used and limiting these uses solely for the purposes of the 2008 Bill.[949]

United States of America

In contrast to the EU, the US has no overarching legislation relating to the protection of privacy; rather, it has taken a sectoral approach to data protection. Nonetheless, since the terrorist attacks of September 11th 2001, a plethora of security enhancing legislation, which have implications for privacy, have been enacted.

The US has two key pieces of legislation in relation to the collection and processing of biometric information: the *Uniting and Strengthening America by Providing Appropriate Tools Required to Intercept and Obstruct Terrorism Act (USA Patriot Act)*[950] and the *Enhanced Border Security and Visa Entry Reform Act.*[951]

The *USA Patriot Act*, which was passed within two months of September 11th, and renewed in March 2006, mandated the development of biometric technology and tamper resistant, machine readable travel documents to assist in the establishment of an entry-exit data system for non-US nationals entering the country. This was followed shortly after by the *Enhanced Border Security and Visa Entry Reform Act*, which made a number of provisions including the introduction of machine readable, tamper resistant, biometric travel documents; the establishment of a data system, which would allow state officials access to law enforcement and intelligence information; the establishment of a visa waiver programme requiring participating countries to issue machine readable, tamper resistant, biometric travel documents; and the requirement for all commercial flights and vessels coming to the US from international ports to provide, by electronic means, information relating to passengers and crews.

949 Irish Human Rights Commission (2008). *Observations on the Immigration, Residence and Protection Bill 2008.* Irish Human Rights Commission, Dublin, 126p.

950 *Uniting and Strengthening America by Providing Appropriate Tools Required to Intercept and Obstruct Terrorism Act (USA Patriot Act)*, Public Law 107–56 (2001).

951 *Enhanced Border Security and Visa Entry Reform Act*, Public Law 107–73 (2002).

Section 303 (c)(1) of the 2002 Act initially[952] stipulated that:

> Not later than October 26, 2004, the government of each country that is designated to participate in the visa waiver program established under section 217 of the Immigration and Nationality Act shall certify, as a condition for designation or continuation of that designation, that it has a program to issue to its nationals machine readable passports that are tamper-resistant and incorporate biometric and document authentication identifiers that comply with applicable biometric and document identifying standards established by the International Civil Aviation Organization.

Following the 2001 attacks, the US introduced a number of other law enforcement measures, including the National Security Entry-Exit Registration System (NSEERS). The NSEERS programme required male non-citizens over the age of 16, from certain countries (predominantly Islamic), to provide fingerprints and photographs when entering the US.[953] NSEERS was intended to act in conjunction with the European Schengen Information System. Another of these measures was the Student Exchange Visitor Information System (SEVIS), which stored information on individual foreign students using Internet-based technology.[954]

In 2003, the US DHS announced that a new programme entitled the United States Visitor and Immigration Status Indicator Technology (US-VISIT) would amalgamate the NSEERS and SEVIS systems. Under the new larger system, non-citizens, entering *via* air, sea or land ports are required to provide fingerprints and photographs and, possibly, iris scans.[955] The aim is to use biometric identifiers to create an "electronic check in/check out system" that can be used not only for reasons of national security, but also for locating visa violators. DHS officials have indicated that they might, depending on technological developments, collect further types of biometric information in the future.[956]

A number of states in the US have either enacted or are in the process of enacting legislation to introduce biometric drivers' licences. While there are no official figures, it has been estimated that only approximately 34 per cent of people over the age of 18 in the US hold passports.[957] Therefore, drivers' licences have become a *de facto* mode of identification, with airlines using them to verify the identity of passengers taking domestic flights. In the US, drivers' licences are issued on a state rather than federal basis, which has led to many different types of licence being issued, thus, making it difficult to visually determine their authenticity. The *REAL ID Act* (2005)[958] established nationwide standards for state issued drivers' licences and other identification cards. REAL ID compliant identification must incorporate a digital face image and

---

952 The deadline initially stipulated was extended by one year.

953 Legomsky SH (2005) The ethnic and religious profiling of noncitizens: national security and human rights. *Boston College Third World Law Journal* 25(1): 161–196.

954 *ibid.*

955 *ibid.*

956 Seghetti LM and Viña SR (2004). *US Visitor and Immigrant Status Indicator Technology Program (US-VISIT).* CRS Report for Congress, Washington, 32p.

957 The Economist (2005). Canada and the United States: the unfriendly border. *The Economist* 376(8441): 31–32.

958 *The Real ID Act*, Public Law 109-13 (2005).

some type of common machine readable technology, such as RFID. Compliance is necessary in order to gain access to federal facilities, boarding commercial flights and entering nuclear power plants. While federal agencies were initially scheduled to accept only drivers' licences and other identification cards that meet REAL ID standards by 11 May 2008, the DHS has granted an extension until the end of December 2009.[959] The final deadline set for all REAL ID compliant state issued drivers' licenses and identification to be issued is 1 December 2017.[960]

In 2001, the US also passed the *Aviation and Transportation Security Act*, which provided for the introduction of passenger and baggage screening capabilities, background checks and screening of aviation employees, fortified cockpit doors and other security measures. The Act also provided for the use of biometric technology in order to identify employees and law enforcement officers entering secure areas of an airport, the use of "voice stress analysis, biometric or other technologies" to prevent a person who may pose a threat from boarding an aircraft and it also called for the consideration of a requirement for all pilot licences to incorporate a photo and biometric information. Section 137 of the Act, which deals with research and development, states that $20 million will be appropriated for research into a number of areas including the development of biometrics for identification and threat assessment.

In May 2008, the Committee on Homeland Security introduced H.R. 5982, *The Biometric Enhancement and Airport-Risk Reduction (BEAR) Act*. This Bill was introduced for the purpose of transportation security and to conduct a study on how airports can convert to uniform, standards-based and interoperable biometric identifier systems for airport workers with access to secure or sterile areas of an airport, as well as other purposes. On 18 June the House of Representatives passed the Bill. It was then referred to the Senate where it was further referred to the Committee on Commerce, Science, and Transportation for more in-depth consideration. At the time of writing, this committee was still considering the *BEAR Act*.

As noted above, there is no explicit right to privacy in the US' Constitution nor is there an independent privacy protection agency.[961] However, the Supreme Court has ruled that there is a limited constitutional right to privacy, based on several provisions in the Bill of Rights. This includes a right to privacy from government surveillance into an area where a person has a "reasonable expectation of privacy".[962] Moreover, some US states have incorporated explicit privacy protections into their constitutions.[963] Currently, privacy laws in the US do not provide the same protection to citizens as the EU. For instance, the *US Privacy Act* of 1974 protects only data held in government databases. The US does not have comprehensive privacy protection legislation covering the private sector. There are a number of laws, nonetheless, which provide protection for some specific categories of personal data – including the *Fair Credit Reporting Act* of 1970, the *Family Educational Rights and Privacy Act* of 1978, the *Privacy Protection Act* of 1980, the *Cable Communications Policy Act* of 1984 and the *Video Privacy Protection Act* of 1988.

---

959 Department of Homeland Security (2008). *DHS Releases REAL ID Regulation*. Press release from the Department of Homeland Security 11 January 2008. Available online at: http://www.dhs.gov/xnews/releases/pr_1200065427422.shtm, accessed 3 November 2008.

960 *ibid.*

961 The Office of Management and Budget plays a limited role in setting policy for federal agencies under the *Privacy Act* (1974).

962 *Katz v. United States*, 386 U.S. 954 (1967).

963 *Constitution of California* (1879) Article 1, Section 1.

As was mentioned previously, Article 25 of the 1995 European Data Protection Directive prevents the transfer of personal data to third countries that do not provide an "adequate level of protection" to European citizens. Consequent to the different approaches taken by Europe (comprehensive) and the US (sectoral) to data protection, a data transfer framework, known as "Safe Harbor" was agreed between the EU Commission and the US Department of Commerce. Safe Harbor is seen as "an important way for US companies to avoid experiencing interruptions in their business dealings with the EU or facing prosecution by European authorities under European privacy laws".[964] A company's decision to participate in Safe Harbor is entirely voluntary. What is required is that the company abides by a number of principles relating to notice, optional participation of citizens, transfer to third parties, access, security, data integrity and enforcement.[965] It should be noted that Safe Harbor is operated on a self-certified, self-regulatory basis and a number of companies were found to have failed to comply fully with its principles.[966,967]

Australia

Australia has an exceptionally high number of migrants (economic and humanitarian) crossing its borders annually. Since 1945, 6.8 million people have migrated to Australia with almost one in four of the current population of 21 million people having been born outside of the country.[968] Due to concerns regarding identity fraud as well as national security, Australia has been active in introducing biometric applications to ensure that only those entitled to cross its borders. The Department of Immigration and Citizenship (DIAC), in conjunction with other government departments and the Office of the Privacy Commissioner, initiated a strategy for identity management know as "Biometrics at the Border". In the 2004–2005 budget, funding worth $214 million over four years was granted for the development of the new scheme.[969] Biometrics at the Border included the introduction of an e-passport for Australian citizens, which contains an RFID microchip storing a digital image of the passport holder. It also allowed for the use of facial images at border control, with fingerprint scans being used for "identity risks [which] are perceived to be greater", for example, refugees and people held in immigration detention centres.[970] Biometrics at the Border will also consist of a database known as the Identity Services Repository (ISR) and airport self-service kiosks known as SmartGate,

---

964 For more information see the Safe Harbor Overview. Available online at: http://www.export.gov/safeharbor/eg_main_018236.asp, accessed 5 November 2008.

965 For more information see the Safe Harbor Overview. Available online at: http://www.export.gov/safeharbor/eg_main_018236.asp, accessed 5 November 2008.

966 Dhont J, Asinari MVP, Poullet Y, Reidenberg JR and Bygrave LA (2004). *Safe Harbour Decision Implementation Study*. European Commission, Brussels, 263p. Available online at: http://ec.europa.eu/justice_home/fsj/privacy/docs/studies/safe-harbour-2004_en.pdf, accessed 5 November 2008.

967 Commission of the European Communities, Commission Staff Working Document (2004). *The implementation of Commission Decision 520/2000/EC on the adequate protection of personal data provided by the Safe Harbour privacy Principles and related Frequently Asked Questions issued by the US Department of Commerce*. Commission of the European Communities, Brussels, 14p. Available online at: http://ec.europa.eu/justice_home/fsj/privacy/docs/adequacy/sec-2004-1323_en.pdf, accessed 5 November 2008.

968 Australian Government, Department of Immigration and Citizenship (2008). *Fact Sheet 2. Key Facts in Immigration*. Available online at: http://www.immi.gov.au/media/fact-sheets/02key.htm, accessed 14 November 2008.

969 Australian Government, Department of Immigration and Citizenship (2007). *Identity Matters: Strategic Plan for Identity Management in DIAC 2007–2010*. Department of Immigration and Citizenship, Canberra, 65p.

970 *ibid.*

which allows e-passport holders[971] to self-process through passport control. The ISR will allow for all information relating to identity, including biometric data on non-citizens, obtained during visa application, onshore compliance of immigration detention actions to be managed and tracked by one system.[972] The Office of the Privacy Commissioner was allocated part of the funding and as a result has committed to undertake three audits per year of key projects in the Biometrics for Border Control programme.[973] The DIAC intends that Biometrics at the Border should be fully operational by June 2009.

In 2004, the *Migration Legislation Amendment (Identification and Authentication) Act*[974] was enacted in order to provide a legislative basis for collecting personal information, including biometric identifiers. According to Section 5(A)(1) personal identifiers are defined as fingerprints, including those taken by digital live scanning technologies, digital photographs, iris scans and audio/video recording. Section 5(A)(3) lists a number of purposes for which the collection of personal identifiers is permitted, including:

- assisting in the identification of, and to authenticate the identity of, any non-citizen;
- improving the integrity of entry programmes, including passenger processing at Australia's border;
- enhancing the Department's ability to identify non-citizens who have a criminal history, who are of character concern or who are of national security concern; and
- combating document and identity fraud in immigration matters.

The Act allows for the establishment of a database containing "indexes of persons who have provided personal identifiers and their identifying information". The Act also contains strict rules in relation to non-"permitted disclosure" of personal information, and a person found guilty of unlawfully disclosing personal data could face imprisonment. It also provides safeguards relating to the storage and management of databases.

The *Privacy Act 1988* and *the Privacy (Private Sector Amendment) Act 2000* provide protection to personal information collected by private and public sector organisations. The 1988 Act sets out a number of "Information Privacy Principles", which cover, for example, the manner and purpose of collection, storage and security, recording, access, uses and disclosure of personal information. Like other data protection legislation, it is unclear whether biometric data is safeguarded. However, in 2008 the Australian Law Reform Commission's (ALRC) privacy review, recommended that biometric information (including biometric templates) should be classified as sensitive information under the Privacy Act in order to ensure that it is afforded a higher level

971 SmartGate is currently available only to Australian and New Zealand e-passport holders over the age of 18 but there are plans to make it available to other nationalities with e-passports.

972 Australian Government, Department of Immigration and Citizenship (2008). *Fact Sheet 84 – Biometric Initiatives*. Available online at: http://www.immi.gov.au/media/fact-sheets/84biometric.htm, accessed 14 November 2008.

973 Australian Government, Office of the Privacy Commissioner (2007). *The Operation of the Privacy Act Annual Report 1 July 2006 – 30 June 2007*. Office of the Privacy Commissioner, Sydney, 167p. Available online at: http://www.privacy.gov.au/publications/07annrep/07annrep.pdf, accessed 14 November 2008.

974 *Migration Legislation Amendment (Identification and Authentication) Act 2004*, No. 2, 2004. An Act to amend the Migration Act 1958, and for related purposes. Available online at: http://www.comlaw.gov.au/ComLaw/Legislation/Act1.nsf/0/0AB6E280C1BF6537CA257421000655FD/$file/0022004.pdf, accessed 12 November 2008.

of protection.[975] It should be noted, however, that the ALRC also recommended that the Privacy Act should remain "technology-neutral" so as to avoid the Act rapidly becoming outdated. According to Professor David Weisbrot, President of the ALRC, "we need privacy principles that are flexible enough to be adapted over time as the technology continues to evolve".[976]

Canada

In Canada a rather different approach to legislating the use of biometrics has been adopted. The Office of the Information and Privacy Commissioner (IPC) of Ontario has collaborated with representatives from the public and private sector to identify and address privacy concerns prior to the implementation of biometric technology, in particular, with respect to welfare fraud. Canada, like many other jurisdictions around the world, has been attempting to combat benefit fraud for some years. One such type of fraud, known as "double dipping", refers to the practice of using multiple identities to obtain welfare assistance illegally. The city of Toronto decided to consider the introduction of biometric technologies to overcome the problem and consulted with the IPC and the Ministry of Community and Social Services in order to develop a legislative framework.

The result of this collaboration was the development by the IPC of procedural and technical safeguards, wherein the IPC made a number of recommendations. The IPC proposed that:

- biometric data should be encrypted;
- the use of encrypted biometric data should be restricted to authentication of eligibility for welfare as opposed to being used as an instrument of social control or surveillance;
- the identifiable finger print cannot be reconstructed from an encrypted finger scan stored in the database;
- strict controls regarding who can access biometric data should be established;
- external agencies wishing to access biometric data *e.g.* the police would have to obtain a court order or warrant in advance; and
- any benefit information should be stored separately from biometric data.[977]

In 1997, the Ontario government passed the *Social Assistance Reform Act*, which incorporated a number of the IPC's recommendations. According to Section 75(3), biometric information must not be disclosed to third parties unless either a warrant or court order is provided.[978] Section 75(6) of the Act states that, "an administrator shall ensure that biometric information collected under this Act is encrypted forthwith after collection, that the original biometric information is destroyed after encryption and that the encrypted biometric information is

975 Australian Law Reform Commission (2008). *For Your Information: Australian Privacy Law and Practice*, Volume 1, Report 108. Australian Law Reform Commission, Sydney, 830p.
Available online at: http://www.austlii.edu.au/au/other/alrc/publications/reports/108/vol_1_full.pdf, accessed 17 November 2008.

976 Australian Law Reform Commission (2008). *Technology-neutral privacy principles should govern rapidly developing ICT*. Press release from the Australian Law Reform Commission published 11 August.
Available online at: http://www.alrc.gov.au/media/2008/mbn2.pdf, accessed 17 November 2008.

977 Cavoukian (1999) *op. cit.*

978 *ibid.*

stored or transmitted only in encrypted form".[979] Section 75(7) prohibits the implementation of a system that can reconstruct or retain the original biometric from encrypted biometric information or that can compare it to a copy or reproduction of biometric information not obtained directly from the individual.[980] The *Social Assistance Reform Act, 1997* was revoked in March 2001; however, the same biometric provisions are listed in the *Ontario Works Act, 1997* which was last amended in 2006.[981]

After September 2001, Canada introduced a number of anti-terrorism initiatives including the establishment of a national fingerprint identification system "LiveScan", which is run by Citizenship and Immigration Canada, Transport Canada and the Royal Canadian Mounted Police (RCMP). The LiveScan machines capture fingerprints, biographical information and photographs of all people seeking refugee status and other people whose identity may be in doubt. In a report by the Auditor General in 2004, the view was expressed that LiveScan was an inadequate system. It was pointed out that there was no information provided regarding the benefits of the system, nor any risk analysis and that a Real Time Identification system for fingerprints had not been implemented. According to the report there was a two and a half month backlog of fingerprint verifications. The Auditor General claimed that the benefits of the system were "marginal at best".[982]

In April 2004, the Canadian government published a strategic framework entitled *Securing an Open Society: Canada's National Security Policy*, which is an action plan designed to ensure that Canada can respond to and anticipate current and future threats. In *Securing an Open Society* the Canadian government has adopted an integrated approach to security issues. One such issue is border security. The action plan lists a number of measures which the government has taken to secure its borders, including the introduction of biometric passports and facial recognition systems, and the rolling out of biometric immigration systems to enhance the design and issuance of travel and identity documents and to verify the identity of travellers at ports of entry.[983]

Canadians are afforded a constitutional right to privacy under the *Canadian Charter of Rights and Freedoms*[984] and their right to privacy is also protected by the *Canadian Human Rights Act*, 1977 as amended. Canada has two pieces of federal legislation relating to privacy, namely the *Privacy Act* of 1982 and the *Personal Information Protection and Electronic Documents Act* (PIPEDA) of 2000. The Privacy Act regulates government departments and agencies by limiting their collection, use and disclosure of personal information. It also allows individuals to access their personal information and to request correction where necessary. PIPEDA regulates how the private sector collects, uses and discloses personal information throughout its commercial

---

979 *ibid.*

980 *ibid.*

981 *Ontario Works Act, 1997*, S.O. 1997, Chapter 25, Schedule A.
Available online at: http://www.e-laws.gov.on.ca/html/statutes/english/elaws_statutes_97o25a_e.htm, accessed 17 November 2008.

982 Office of the Auditor General of Canada (2004). *2004 March Report of the Auditor General of Canada.* Office of the Auditor General of Canada, Ontario.
Available online at: http://www.oag-bvg.gc.ca/internet/docs/20040303ce.pdf, accessed 17 November 2008.

983 Privy Council Office Canada (2004). *Securing an Open Society: Canada's National Security Policy.* Privy Council Office Canada. Ontario.

984 *Dagg v. Canada (Minister of Finance)* [1997] 2 S.C.R. 403.

activities. When first enacted, PIPEDA applied only to federally regulated organisations, such as banks and airlines but now also applies to other sectors, for example, retail, manufacturing and the service industry. The Office of the Privacy Commissioner of Canada (OPC) has strongly advocated for a reform of the Privacy Act so that it would take account of new challenges facing privacy, for example, the Internet, electronic surveillance, global positioning systems and biometric technologies. Indeed, in its annual report 2004–2005, the OPC stated that "characterizing the current Act as dated in coping with today's realities is an understatement – the Act is tantamount to a cart horse struggling to keep up with technologies approaching warp speed".[985] In June 2006, the OPC presented a plan for reforming the Privacy Act to the House of Commons' Standing Committee on Access to Information, Privacy and Ethics, which contained a number of recommendations including a requirement for government institutions to assess the privacy impact of programmes or systems prior to their initiation, and to publicly report the results of assessments and the need for privacy management of transborder data flows.[986] In April 2008, the Standing Committee on Access to Information, Privacy and Ethics agreed to conduct a review of the Canadian Privacy Act.[987]

The Council recommends that biometric data should be classified as sensitive personal information and as such afforded greater protection. Consequently, the Council is of the opinion that Ireland's existing data protection legislation does not deal sufficiently with the privacy concerns presented by the increasingly mainstream use of biometrics. The Council welcomes the decision by the Minister for Justice, Equality and Law Reform in November 2008 to establish a committee to review current legislation[988] and urges the committee to consider the privacy/data protection implications arising from biometric technologies.

985 The Privacy Commissioner of Canada (2005). *Annual Report to Parliament 2004–2005. Report on the Privacy Act.* The Privacy Commissioner of Canada, Ontario, 81p. Available online at: http://www.privcom.gc.ca/information/ar/200405/200405_pa_e.asp, accessed 17 November 2008.

986 The Privacy Commissioner of Canada (2006). *Government Accountability for Personal Information: Reforming the Privacy Act.* The Privacy Commissioner of Canada, Ontario.
Available online at: http://www.priv.gc.ca/information/pub/pa_reform_060605_e.cfm, accessed 17 November 2008.

987 Standing Committee on Access to Information, Privacy and Ethics (2008). *Minutes of Proceedings.* Available online at: http://www2.parl.gc.ca/HousePublications/Publication.aspx?DocId=3426491&Language=E&mode=1&Parl=39&Ses=2, accessed 17 November 2008.

988 de Bréadún D (2008). Ahern sets up group to review data protection law. *The Irish Times*, 1 November 2008.
Available online at: http://www.irishtimes.com/newspaper/ireland/2008/1101/1225321623499_pf.html, accessed 14 November 2008.

APPENDICES

Appendix A:
Report from the Focus Groups

It is the established policy of the Irish Council for Bioethics to conduct some form of public consultation relating to the topic the Council is currently working on, *i.e.* biometrics in this case. However, as biometrics is an emerging and complex set of technologies, the Council took the view that the level of knowledge of these technologies and the issues associated with them among the general public would be somewhat limited. Focus groups were considered to be the optimal mechanism to elicit the full range of ideas, attitudes, experiences and opinions held by a selected sample of respondents on the complex topic of biometric technologies. This methodology allowed for clarification, probing and follow up questions, which would not have been possible to the same extent in a written questionnaire. The Council engaged with the market research company Red C to facilitate and organise the focus groups. In addition, the Council produced an information leaflet (see Appendix B) highlighting some of the main ethical issues associated with biometrics. The purpose of the leaflet was to help inform the focus group discussions and each participant received a copy of the information leaflet prior to attending a focus group.

Similarly to previous consultations the Council has conducted, the purpose of this exercise was to provide an opportunity for the general public to highlight any issues or concerns they might have with regard to biometric technologies and their associated applications. Any issues raised during the focus groups were then incorporated into the Council's own deliberations on this topic. The focus groups therefore assisted the Council by providing a different perspective on the ethical and social issues associated with biometrics that the Council were already considering. The focus groups were made up of a very small number of people who voluntarily participated and it cannot be assumed that their views and perceptions are representative of the wider public. Moreover, it was never the intention of the Council that the results of the focus groups would act as a substitute for a detailed, quantitative survey of public opinion on biometrics.

C

Biometrics Qualitative Research March 2008

RESEARCH EVALUATION DIRECTION CLARITY

Overview of Research Objectives

- **The primary objective of the research was to explore public opinion and reaction to the concept and application of biometrics.**

- More specifically, the research would provide insight into the following:
  - *general thoughts about biometrics among the general public;*
  - *awareness, understanding and experience of biometrics;*
  - *positives and negatives associated with its various applications;*
  - *specific concerns and worries associated with biometrics and its different applications;*
  - *highlighting specific information gaps.*

Methodology and Sample Design

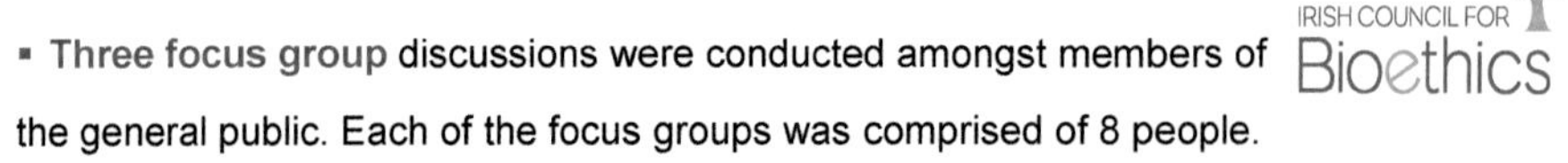

- **Three focus group** discussions were conducted amongst members of the general public. Each of the focus groups was comprised of 8 people.
- Groups were spread across a **broad spectrum** of demographics as follows:

DUBLIN	MULLINGAR
20-29 years, BC1,* Male and Female	45-60 years, C1C2, Male and Female
30-44, C2DE, Male and Female	

- **Pre-task:** Respondents were provided with a Biometrics information leaflet (see Appendix B) to read in detail prior to attending the group discussions.
- **Focus groups** were conducted on 26 and the 28 March 2008.
- All groups were **moderated** by Sandra Reape, Associate Director at RED C Research and was attended by a member of the Irish Council for Bioethics Secretariat.

**For the purposes of the focus groups the following standard social demographic categories were used:*

B = Middle Class; C1 = Lower Middle Class; C2 = Skilled Working Class; D = Other Working Class; and E = Lowest Levels

Overall Reaction to Biometrics

- Information leaflet and concept of biometrics received **very high levels of interest** and enthusiasm across all groups *(ages, genders and social grades).*
- Evident **desire to understand and learn** more about the concept of and its applications.
- **General acceptance** of such developments … it feels like a natural progression from where we are today … constant evolution of current systems and processes

Signature  *Chip and Pin*  *Fingerprint?*

- Overall, there were **few significant worries and concerns** … information and reassurance around the capture, storage and usage of biometric information was of key importance.

redc

Awareness and Experience of Biometrics (i)

- All were **aware of "biometric" applications** in some form or other, be it "fingerprinting", "iris scanning" etc. from overseas travel, work environments etc.
- The term **"biometrics"** was known only to a **small few** but they did not make the link to the actual applications:
- *"I've heard of biometrics and of fingerprinting, scanning etc. but I didn't know the link between the two!"*
- Many **had experience** of some biometric applications both in Ireland and overseas.

redc

Awareness and Experience of Biometrics (ii)

"My parents had to scan their fingers to get access to theme parks in Disney World."

"I've scanned my finger to get access to a casino in town."

"We had to scan our hand to get access to the Children's ward or the morgue."

"We're getting a new clock-in system - using our finger instead of a clock-in card."

redc

Awareness and Experience of Biometrics (iii)

- **TV programmes** such as *CSI* and movies have done a lot to raise the profile of biometrics … means we are less shocked by new developments…
- While biometric applications in movies may be "far fetched"… sense that we will see some variation or application of them in the future!

  "Sure James Bond was using fingerprints and iris scanning 20 years ago!"

- Evidence that males somewhat more informed … generally more likely to watch science-based programmes.

redc

Passports/International Travel (i)

- Respondents **very well travelled with a lot of experience** with developments and procedures in other countries.
- **Heightened sense of awareness** of restrictions and requirements for entering various countries (post 911 and other recent terrorist events). Respondents cited:
- *USA* - Finger printing and iris scanning.
  Spain - Requirement to send your details to Spanish authorities in advance of travel.
- General **acceptance and understanding** of need for new requirements/ added security measures in the interests of national and personal security.

  Questions posed included:

- *Will finger printing and eye scanning become the norm?*
- *Will a couple of methods be used as well as digital photo/chip?*
- *What other information will be on the chip?*

redc

Passports/International Travel (ii)

- Many had new biometric passports and understood the reasons and benefits behind the new formats:
  - *Chip seen to hold a digital version of your photo and personal details*
  - *Seen to protect our identity – changing the picture doesn't mean they can take my identity… there are multiple checks in place!*
- Many respondents were quite au fait with the concept of digital photos for passports…based on facial dimensions not photo only.

"measures the distance from eye to nose, nose to mouth etc."
"some kind of chip with your details on it"

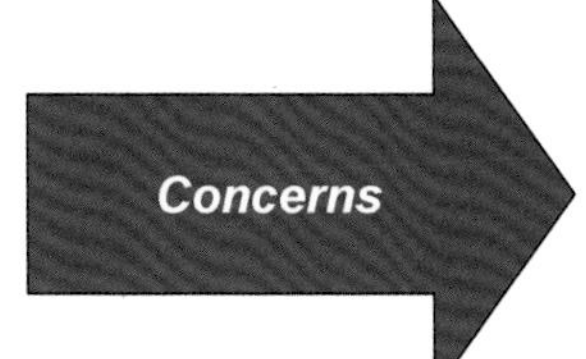

- **What happens if I have an accident/facial surgery?**
- *Do I need to get a new passport?*
- *Is the onus on me to inform the passport office?*
- *Would I just take medical records and proof with me when I travel?*

Voice Recognition (i)

- Can imagine the speed of access and **greater convenience** for the likes of telephone banking.

"I'd prefer that to punching in endless numbers."

- **Vague understanding** among many about how the system would operate

However, as anticipated, only a very small minority had any understanding of the more technical details around the authenticity of personal sound waves, etc.

- General assumption that voice recognition would be used in conjunction with **another layer of ID**… such as a PIN number.

redc

Voice Recognition (ii)

Concerns

- Overriding fear is voice replication or imitation by another person or by technology:

  "Someone could mimic my voice."
  "They can do all sorts with computers these days…"

- What happens if my voice (tone or pitch) changes? What if I have a cold, a different accent etc.
- Are our voices really unique?

  "Sure my mother's sounds exactly like I do… we always get confused on the phone!"

redc

Time and Attendance - School

- Using biometrics as a means of tracking child attendance at school was **highly regarded** by all:
  - *Our prize possessions… we will do anything to ensure they are fully protected. "Great to know where they are at all times."*
  - *The rise in school crime has made this an ever increasing issue.*
  - *A system such as "fingerprint scanning" overcomes the problem of kids clocking in for each other…falsifying attendance and associated issues.*
- Evidence that some forms of digital/computerised attendance monitoring is **currently in place** in many schools:
  - *Many had personal experience - themselves, kids, grandkids..who had smart cards to register in school in both Dublin and Mullingar.*
- The idea of **automatic alerts** (text/email etc.) to the parents was seen as a good idea but in need of some controls over when sent.
  "What if my child was running late because the bus was delayed??"

redc

Time and Attendance – Work (i)

- Some **had experience** of 'biometric' applications to monitor attendance at work – hospitals, supermarkets, factory floors etc.
- General feeling that we have been using swipe/pass cards for entry to buildings and clocking in for years… this feels like a **natural progression**.
  - *Ease and convenience: "You won't forget your finger!"*
  - *Perception that it's more effective and more accurate read of attendance.*
  - *Individual responsibility for clocking in… you can't be asked by colleagues to do it for you…. More equitable system.*
- Seen as ok if such systems are used to register attendance and working hours….but attempts to **overly monitor and control personal movements** within the workplace are **rejected.**

Time and Attendance – Work (ii)

- Some factory workers reported usage of systems that **monitor every movement**…*e.g. tracking time away from production lines and calling people via intercoms to come back to the line etc…*
  - *Embarrassing, intrusive and invasion of personal freedom and space.*
- What happens if the **system fails**…e.g. doesn't register my fingerprints and I don't get pay for 'x' days
  "How do I prove I was there?"
- Respondents requested **specific reasons** to clarify why biometric systems are being introduced
  - *What are the company reasons and benefits over more traditional methods?*
- General consensus that this more **"scientific"** approach would be **more accurate and safer**.

Physical and Health Concerns

- Overall, respondents' **health concerns** regarding biometric applications were **minimal**.
- In terms of **sanitation**... hand-scanning machines were seen as no dirtier than ATM machines:

  "I think some machines have a built in cleaning device after each scan"

- Somewhat more concerned about **cleanliness** of putting their head into a hold for iris scanning.
  - But many assume such scanning could be done from a distance.
    "Surely you could just stand in a spot and they could scan your eyes from a distance??"
- How **far away is the scanner/laser** from your eyes...
  "What damage could it do?" "Would repeated exposure be harmful?"
- Would body scans be harmful if I was **pregnant**?

redc

Freedom of Information (i)

- General feeling that a lot of information about us is **already shared by different departments** and used for purposes other than those originally provided for:

  - *Supermarkets offering **loyalty cards** use our purchase behaviour to target us with products.*
  - ***Electoral register** is used to send us circulars, junk mail etc.*
  - ***Revenue and Criminal Assets Bureau** seen to have widespread access to our personal information and movements....they can find out anything about you!*

    "They have access to my passport details, they can track where I have been and can ask me how I paid for it all!!"

- Belief that **government departments** would readily share information about you with other departments.

redc

Freedom of Information (ii)

- The majority said they were **ok with having information** about them made available to different departments

"It doesn't matter to someone like me… I have nothing to hide"

- Happy to have information shared as long as the information provided to different bodies/department is **relevant to the needs** of that department and what it needs to know about me!

E.g. the library don't need to know my blood group…but ok for hospital to have my health insurance details…etc.

redc

Security (i)

- **Widespread awareness** across age groups of recent breaches in security in the UK and the Blood Transfusion Board.

Heightened awareness and consideration around security of personal information and details.

- **ATM** skimming a concern for most… concept understood and a real reality.
- Many understood the concept of **'Encryption'** even many older respondents…. *Feeling that data in many instances cannot be interfered with even if someone gets access to it!*

- While **security of personal information** is a concern for all at some level it is **not top of mind** for most.

"What could they really do with my blood information if they got it??"
"Sure it's encrypted what could they do?"

redc

Security (ii)

- ***A general trust in science and scientific approaches to personal security and protection was evident across groups.***
- Perception and expectation that scientific means of identification add a **deeper layer of security**;
  - *Unique and individual – fingerprint, eye, DNA.*
  - *Harder (but not impossible) to replicate.*
  - *More accurate – this is science!*

- All aware that fingerprints can be copied.
  "Sure you could get them off that glass there, or the table.."
- This creates a doubt about fingerprinting..would need an additional layer of identification!
- Experience and awareness of recent and ongoing security breaches makes us sceptical:
- *"There is always someone who can out wit the system"*

redc

Security (iii)

- General feeling that DNA and IRIS were **more difficult** or impossible to replicate.
- For many (rightly or wrongly) using **DNA** was seen as the **ultimate** form of security.
- Despite a significant trust in science as a means of identity protection **doubt and scepticism still exists.**

We would still feel safer with multiple layers of security….
Use a PIN as well!

redc

Information Storage and Protection

IRISH COUNCIL FOR Bioethics

Questions Posed by the Focus Groups

- ***Will there be a central bank of Biometric information?***
  - ***Where** is it held and in what format?*
  - ***Who has access** to my information and under what circumstances?*
  - ***Which organisations/offices/departments** have unlimited access rights and which have restricted access?*
  - *What **power do I have** in controlling who has access to my information? "can I select/tick box who I want to access my information?"*
  - *What are **my rights**? Who can I trust with my information?*
  - ***Who** (body/department) **monitors** the release and maintenance of my information? Is it centrally managed and controlled?*

- ***Assumption that there would be checks in place as the detail and the complexity of information stored about us increases in the future.***

What is the role of the Data Protection Commissioner?

- Assumption for many that the **Data Protection Commissioner** and department would be responsible for **monitoring and controlling information held.**
- And ensuring that any information provided to others about us is done in **a lawful** and fair manner.
- Would expect a **more visible and prominent** role for the Data Protection Commissioner in the future in light of such developments.

Needs to be seen as a real and powerful force in its role as information protector.

redc

Benefits of Biometrics (i)

THE OVERRIDING BENEFIT IS ENHANCED SECURITY AND IDENTITY PROTECTION

National Security

- Biometric means of identification is seen as a more effective means of monitoring the movement of people between countries.
- Allows better and more effective screening of those entering the country.

Personal Security

- Offers higher levels of protection for my personal information such as bank /finance details.
- Allows for more better monitoring of those who are more vulnerable and need to be protected*...children and older people.*

Identity Protection

- Identify proof based on scientific elements seen to afford greater protection – *reduced likelihood that identity can be stolen... or indeed much hard to do it!*

"much better than just having a photo only!"

redc

Benefits of Biometrics (ii)

Greater efficiency/ convenience

- Biometric process offer a speedier and more efficient process

Faster clock in process, easier to access accounts...

Easier to scan finger/eye than detailed document checks....

Equitable system

- Everyone monitored in the same fashion
- Onus is on the person to clock in themselves...

No one under pressure to do it for them or take a risk on behalf of others...

redc

Information and Communications: *Addressing Public Concerns*

Provide detailed information.

Reassure and instil confidence in new system and different biometric applications.

Provide peace of mind that information is being monitored and controlled.

- **What** information is stored and **why** is it needed?
- ***Who*** *will have access to my information and for what* ***purposes****?*
- ***Where*** *will my information be* ***stored*** *and who will monitor and protect it in terms of content and dissemination?*
- ***How and when*** *should information be* ***updated*** *and who is the onus on to do so?*
- *What happens if my* ***features/characteristics*** *change or are altered?*
- *What say do I have in the* ***provision and monitoring*** *of my information?*
- *Will biometrics be used in conjunction with* ***other layers of security****?*

redc

In Summary…

The Council's hypothesis going into the research was that general awareness, experience and understanding of biometrics and its applications would be quite limited, confusing and would generate a number of concerns and worries for future roll-out and development.

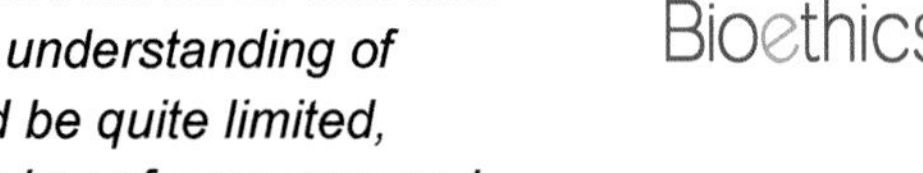

- ***The reality was that people were…***
  - *Very interested and enthusiastic about the concept.*
  - *Quite aware of biometric applications… albeit not the actual term.*
  - *Had personal experience of different applications across a range of situations – travel, work, leisure, school etc.*
  - *General acceptance of developments in this space…natural progression and evolution of current systems and processes.*
  - *Despite a certain trust in science…some scepticism exists around replication and imitation… multiple forms of ID might be more useful.*
  - *A key concern is around the monitoring, control and release of personal information and biometrics to interested parties, companies, government etc.*

Appendix B: Biometrics Information Leaflet

Q1 What are biometrics?

A biometric is any physical or biological feature that can be measured and used for the purpose of identification. Features can be either physiological *e.g.* fingerprint, hand geometry (shape), the face, the iris, the retina or behavioural *e.g.* voice pattern and gait (way of walking).

Q2 For what purposes are biometrics used?

Some biometric features, like fingerprints are considered to be unique to each individual and, in general, biometrics, such as iris pattern and hand geometry are believed not to significantly change with age. This makes them very useful for identification purposes. As a result, biometrics are used to confirm that individuals are who they say they are, to help identify unknown people or to screen people against a specific watchlist, such as a criminal database.

In response to terrorist and criminal activity, biometrics have been introduced with the aim of improving national security. Biometrics have also been introduced, in order to deal with the migration of millions of people between countries both legally and illegally. For instance, the European Union (EU) EURODAC system was set up to combat the flow of illegal immigrants into Member States. EURODAC is a central European database of fingerprints, which allows a Member State to check wether asylum seekers have previously sought asylum from another European country or whether they have tried to enter the EU illegally. In the US the Visitor and Immigrant Status Indicator Technology (US VISIT) Programme was started following the events of September 11th 2001. Under this system all visitors to the US must have both their index fingers scanned and have their photo taken. This information is then checked against a database of known criminals and suspected terrorists.

Apart from national security, biometrics are also used for several other purposes, such as security and surveillance, law enforcement, e-Commerce (buying and selling online), e-Government (electronic communication between governments and citizens) as well as gaining physical and electronic access to buildings or computer files. Biometrics are also being promoted as a possible solution to identity theft and fraud and are increasingly being used by banks and other commercial organisations to correctly confirm the identity of their customers. (For a list of biometric applications see Table 1).

Table 1

APPLICATION	FUNCTION
Time and Attendance	Individual institutions *e.g.* schools and employers use this system to record attendance and to control who has access to buildings or areas.
Pay By Touch	A commercial tool in the US,which allows consumers to pay for products in supermarkets and pharmacies by scanning their fingerprints and punching in a personal identification number (PIN), which is linked to their credit or debit cards.
Voice Recognition	Software used in telephone banking to create a profile of how a person's voice should sound regardless of what is being said.
Machine Readable Passports	These include a small chip storing a digital version of the passport photo as well as other information relating to name, address, date of birth *etc.* Passports can be used along with face recognition software to confirm the passport holder's identity.
Radio Frequency Identification (RFID)	This involves the same technology as that used for machine readable passports. Microchips are implanted in order to track goods, identify pets, and keep track of people *e.g.* nursing home residents or employees.
Iris Recognition	Provides a method of accessing secure areas *e.g.* border control and, in some airlines, allows pilots access to aeroplane cockpits.

Q3 How are biometrics collected?

In general, biometrics are collected using sensors *e.g.* cameras (face recognition), telephones (voice recognition) and fingerprint scanners. People have to enrol before they can use biometric systems. Enrolment involves a copy of a person's biometric feature being taken, converted into a digital format and stored on an electronic database. For example, in the case of a system that uses fingerprints to grant access to a building, a person would have to have his/her fingerprints taken by a sensor and recorded so that s/he can be recognised in the future. The next time the individual presents his/her fingerprint to the sensor this data is compared to the stored copy (also known as a template) using a mathematical formula. If the templates match the individual is granted access.

Q4 How accurate are biometrics?

While biometrics are said to have many benefits for society, it should be noted that these new technologies are not 100% accurate or secure. For example, some biometrics, such as hand geometry or gait are not totally unique to each individual, therefore, their use is limited to small scale systems *e.g.* a small business might use hand geometry to grant employees access to a building or to monitor employee time and attendance. In addition, while fingerprints are considered to be unique, they can become damaged, which can cause problems with fingerprint scanning systems. Indeed, even if machines are functioning optimally, human error can result in inaccurate data being collected.

Opponents of biometrics have raised concerns about the ability of people to outwit biometric technologies, a practice known as "spoofing". For example, spoofing might involve using fake fingers and fingerprints or contact lenses to fool biometric sensors. However, proponents argue that improvements to the technology such as "liveness" detection *e.g.* detecting the oxygen levels in the blood in fingers or muscle movement in the iris, thereby indicating a person is alive, help to reduce the likelihood of spoofing. They also state that human supervision of biometric systems *e.g.* by airport security or border control officials will help to establish that the biometric used really does belong to the person trying to use the system.

Multimodal biometric systems combine a number of biometrics to identify people *e.g.* fingerprint and facial recognition, which may need to be presented in sequence, at the same time or alternately. Proponents argue that using a combination of biometric identifiers can reduce the potential for spoofing systems, therefore making them more secure. They also argue that multimodal biometrics can provide an opportunity for individuals who cannot enrol with one particular biometric feature *e.g.* due to disability, to enrol using alternative features.

Q5 Are there any health risks related to using biometric technologies?

Concerns have been raised regarding the possible health risks of biometrics. Opponents argue that there is a possibility of eye damage arising from the use of iris scanning equipment. It should be noted that, to date, there have been no reported injuries from using iris scanners. Opponents of biometrics also express unease regarding the cleanliness of the sensors used, stating that participants may feel uncomfortable about touching a hand-geometry scanner or placing their face against an iris-scanner after other people have done so. Proponents of biometrics, on the other hand, claim that safety concerns are unfounded and declare that, for instance, hand-geometry scanners are no more unhygienic than a door handle. In addition, they state that hygiene mechanisms, such as regular cleaning or sterilisation of sensors should invalidate such concerns. Furthermore, proponents argue that the introduction of remote scanning (scanning from a distance) will mean that it will not be necessary to actually touch some types of sensors in order to be identified.

Q6 Are biometrics a threat to privacy?

The right to privacy *i.e.* our right to control access to ourselves and to our personal information is protected by the Irish Constitution. However, this right is not absolute and may be overridden in the interest of society and community safety, as in the case of terrorism or a national emergency *e.g.* the outbreak of a contagious disease, such as human avian influenza (bird flu).

Opponents of biometrics have raised concerns regarding the way in which personal information is obtained, stored, compared with watchlists, and possibly linked to other information about an individual. Critics argue that because biometrics are strongly linked to a person's identity and cannot be changed or reissued in the same way as a password or PIN, there is a serious threat to privacy if such information were to get into the wrong hands. Opponents also state that, depending on the biometric identifier used, additional information can be discovered *e.g.* iris scans can indicate alcohol and drug use and facial biometrics can reveal information, such as sex, age and race. They also express unease regarding the possible use of DNA as a biometric identifier beyond its current use in criminal investigations, which could potentially provide genetic and medical information about an individual and result in their being discriminated against.

Concerns have also been raised that the use of personal information could gradually expand and information could be transferred to third parties with or without a person's knowledge. This phenomenon is known as "function creep". Opponents express unease regarding the risks associated with function creep, where information might be used in ways that were not originally intended *e.g.* marketing or for deciding whether somebody should be offered a job. Opponents also raise concerns in relation to how access to biometric databases will be controlled, and the possibility of unauthorised third parties hacking into systems in order to obtain an individual's personal information.

However, proponents of the technology argue that a lot of personal information can already be collected legally from everyday activities, which can provide information about a person's interests and background and that biometrics pose no greater threat. For example, details of the phone numbers a person calls can be stored, banking and credit card transactions can be tracked, library records can be documented and even a supermarket/shop loyalty card can keep track of an individual's purchases.

Furthermore, proponents argue that using biometrics will enhance privacy and protect personal information from unwanted intrusion. They state that, privacy can be maintained in some instances, by storing biometric information on a smart card, which is carried by an individual as opposed to being held on a central database; by using encryption (putting data into a secret code so it is unreadable by unauthorised people) or by storing biometric templates in a separate database to the database storing personal information *e.g.* name, address and date of birth.

Q7 Is consent necessary for participation in a biometric system?

In order to protect individual autonomy *i.e.* one's ability to make independent choices without any external influences, it has been argued that participation in biometric programmes should be optional. However, this freedom of choice is not always guaranteed *e.g.* as part of the UK's biometric immigration system, an individual coming from outside the EU wishing to obtain a visa must provide 10 fingerprints and have his/her photo taken. In this case, while the system at first appears optional and individuals are asked for their consent, failing to give it effectively means that they will be unable to obtain a visa to travel to the UK. Opponents of biometric technologies state that those individuals who are unwilling to provide biometric data should be given an alternative method of providing the required identification information.

Opponents of biometrics have also expressed unease regarding the use of biometrics for identification purposes without the consent of individuals, an issue that may arise more with the increasing use of remote scanning. They argue that given the personal nature of biometric information, and its association with an individual's identity, consent should always be obtained.

Proponents, on the other hand, argue that consent is not always necessary and that in some instances protecting the greater good of society should outweigh individual autonomy *e.g.* in the case of illegal immigration or international terrorism.

Q8 Will biometric technologies lead to discrimination?

Opponents have raised concerns that some of the information collected from biometric systems could be used in profiling and categorising people, potentially leading to the discrimination of certain groups or individuals. From a security point of view, proponents argue that the use of such profiling techniques has been successful in preventing potential terrorist attacks. However, opponents point out that the presence of biometric security measures will not necessarily prevent future terrorist attacks. For instance, a number of the terrorists involved in the attacks on September 11th 2001 travelled to the US using their own passports and held valid visas.

Q9 Is the introduction of biometrics a proportionate response to possible security threats?

While biometrics are expensive and are less than 100% accurate, many governments believe that the benefits of biometric systems, in terms of national security, greatly outweigh any risks to privacy or health and have decided to introduce a number of programmes *e.g.* biometric passports, immigration programmes and national security systems. Similarly, biometrics have begun to appear in smaller settings, for instance, there are a number of examples in Ireland and abroad where institutions *e.g.* schools and employers have introduced biometric systems in order to record the time and attendance of pupils and staff and some commercial organisations are using biometrics to confirm their customers' identities. Proponents argue that biometric programmes increase security and efficiency and provide greater convenience and peace of mind to consumers.

Opponents of biometrics, however, argue that a "Big Brother" style society will be the inevitable consequence of the increased use of biometric technologies, enhanced surveillance methods and increasingly connected databases. They also raise concerns regarding the appropriateness of introducing expensive, technically complex biometric systems to achieve goals that could potentially be achieved in other ways.

Similar arguments have been made regarding the storage of large amounts of personal and biometric information in central databases, given the potential risks of unauthorised access or the loss of important and sensitive personal records. The potential for such risks was highlighted in the UK, when in 2007 discs containing the personal records of 25 million individuals, including their dates of birth, addresses, bank accounts and national insurance numbers were lost in the post.

Appendix C: Submissions Sought by the Irish Council for Bioethics

The following is a list of the organisations and individuals from which the Irish Council for Bioethics sought submissions:

Professor Tom Coffey (Director, Data Communications Security Laboratory, Department of Electronic and Computer Engineering, University of Limerick)

Data Protection Commissioner

Department of Communications, Energy and Natural Resources

Department of Foreign Affairs

Digital Rights Ireland Ltd

Forensic Science Laboratory

Garda National Immigration Bureau

Information Commissioner

Irish Council for Civil Liberties

Irish Naturalisation and Immigration Service

Information Society Commission

Law Reform Commission

Mr John McDonald (Computer Vision and Imaging Laboratory, Department of Computer Science, National University Ireland Maynooth)

Mr Eugene McKenna (Smart Card Operations Research Enterprise Group, Waterford Institute of Technology)

Dr John McKenna (School of Computing, Dublin City University)

Professor Richard Reilly (Director, Neural Engineering Group, Electronic and Electrical Engineering Department, Trinity College Dublin)

Appendix D: Submissions Received by the Irish Council for Bioethics

The following is a list of the organisations and individuals from which the Irish Council for Bioethics received oral and/or written submissions:

Professor Tom Coffey (Director, Data Communications Security Laboratory, Department of Electronic and Computer Engineering, University of Limerick)

Data Protection Commissioner

Forensic Science Laboratory

Garda National Immigration Bureau

Information Commissioner

Irish Naturalisation and Immigration Service

Law Reform Commission

Mr Eugene McKenna (Smart Card Operations Research Enterprise Group, Waterford Institute of Technology)

Professor Richard Reilly (Director, Neural Engineering Group, Electronic and Electrical Engineering Department, Trinity College Dublin)

Rapporteur Group on Biometrics

Professor Alan Donnelly, Department of Physical Education and Sport Sciences

Mr Stephen McMahon, Irish Patients' Association

Mr Asim A. Sheikh BL, Forensic and Legal Medicine, School of Medicine and Medical Science, University College Dublin

Terms of Reference

1. To consider the ethical, social and legal implications arising from the application of biometric technologies and the collection, use and storage of biometric information.
2. To seek the views of stakeholders and to evaluate the public's perception of issues relating to biometrics.
3. To report on all aspects of the Council's deliberations and conclusions.

The Irish Council for Bioethics

Dr Dolores Dooley, Philosopher and Lecturer in Bioethics, **Chairperson**

Professor Alan Donnelly, Department of Physical Education and Sport Sciences, University of Limerick

Professor Andrew Green, School of Medicine and Medical Science, University College Dublin; National Centre for Medical Genetics, Our Lady's Hospital for Sick Children, Crumlin

Dr Mary Henry, Retired Medical Practitioner and Former Independent Member of Seanad Éireann

Professor Linda Hogan, Irish School of Ecumenics, Trinity College Dublin (until December 2008)

Dr Richard Hull, Department of Philosophy, National University of Ireland, Galway

Dr Peter McKenna, Rotunda Hospital, **Vice Chair**

Professor John Vincent McLoughlin, Department of Physiology, Trinity College Dublin

Mr Stephen McMahon, Irish Patients' Association

Professor Cliona O'Farrelly, School of Biochemistry and Immunology, Trinity College Dublin

Mr Turlough O'Donnell SC, Practising Barrister and Former Chair of the Bar Council of Ireland

Dr Darina O'Flanagan, Health Protection Surveillance Centre (until December 2008)

Professor Richard O'Kennedy, School of Biotechnology, Dublin City University

Mr Asim A. Sheikh BL, Forensic and Legal Medicine, School of Medicine and Medical Science, University College Dublin, **Vice Chair**

Professor David Smith, Royal College of Surgeons in Ireland

Dr Sheila Willis, Director, Forensic Science Laboratory

Secretariat

Dr Siobhán O'Sullivan

Ms Emily de Grae

Mr Paul Ivory

Ms Emma Clancy

Terms of Reference

1. To identify and interpret the ethical questions raised by biomedicine in order to respond to, and anticipate, questions of substantive concern.
2. To investigate and report on such questions in the interests of promoting public understanding, informed discussion and education.
3. In light of the outcome of its work, to stimulate discussion through conferences, workshops, lectures, published reports and where appropriate to suggest guidelines.

Abbreviations

AFIS: automated fingerprint identification system

AIDS: acquired immune deficiency syndrome

ALRC: Australian Law Reform Commission

ATM: automated teller machine

BEAR Act: *Biometric Enhancement and Airport-Risk Reduction Act*

CCTV: closed circuit television

DHS: Department of Homeland Security (US)

DIAC: Department of Immigration and Citizenship (Australia)

DNA: deoxyribonucleic acid

DSFA: Department of Social and Family Affairs (Ireland)

ECG: electrocardiogram

ECHR: European Court of Human Rights

EDPS: European Data Protection Supervisor

EEG: electroencephalogram

EER: equal error rate

EU: European Union

FAR: false accept rate

FBI: Federal Bureau of Investigation

FMR: false match rate

FNMR: false non-match rate

FTA: failure to acquire rate

FTE: failure to enrol rate

FRR: false reject rate

ICAO: International Civil Aviation Organization

ICJ: International Commission of Jurists

ID: identity or identification

INSPASS: US Immigration and Naturalization Service's Passenger Accelerated Service System

IPC: Office of the Information and Privacy Commissioner of Ontario

IRIS: Iris Recognition Immigration System

ISR: Identity Services Repository

NSEERS: National Security Entry-Exit Registration System (US)

NSTC: National Science and Technology Council (US)

OECD: Organisation for Economic Co-operation and Development

OPC: Office of the Privacy Commissioner of Canada

PIN: personal identification number

PIPEDA: *Personal Information Protection and Electronic Documents Act*

PPS number: Personal Public Service number (Ireland)

RFID: radio frequency identification

ROC curve: receiver operating characteristics curve

SAIVOX: Sistema Automático de Identificación por Voz

SEVIS: Student Exchange Visitor Information System (US)

SNP: single nucleotide polymorphism

3D: three dimensional

2D: two dimensional

UAE: United Arab Emirates

UK: United Kingdom

UN: United Nations

US: United States of America

US-VISIT: US Visitor and Immigrant Status Indicator Technology

WP29: Article 29 Working Party

Legal Instruments and Regulations

National

Australia

Privacy Act (1988)

Privacy (Private Sector Amendment) Act (2000)

Migration Legislation Amendment (Identification and Authentication) Act (2004)

Austria

Data Protection Act (1978)

Canada

Canadian Human Rights Act (1977)

Canadian Charter of Rights and Freedoms (1982)

Privacy Act (1982)

Ontario Works Act (1997)

Social Assistance Reform Act (1997)

Personal Information Protection and Electronic Documents Act (PIPEDA) (2000)

Denmark

Private Registers Act (1978)

Public Authorities' Registers Act (1978)

France

Act n°78-17 on Data Processing, Data Files and Individual Liberties (1978)

Germany

Hesse Data Protection Act (1970)

Federal Data Protection Act (1977)

Ireland

Bunreacht na hÉireann, The Irish Constitution (1937)

Data Protection Act (1988)

Data Protection (Amendment) Act (2003)

Social Welfare (Consolidation) Act (1993)

Education (Welfare) Act (2000)

European Convention on Human Rights Act (2003)

Social Welfare (Consolidation) Act (2005)

Social Welfare and Pensions Act (2007)

Passport Act (2008)

Immigration, Residence and Protection Bill (2008)

Criminal Justice (Surveillance) Act (2009)

Sweden

The Data Act (1973)

United Kingdom (England)

Justices of the Peace Act (1361)

United States

Constitution of California (1879)

Fair Credit Reporting Act (1970)

Privacy Act (1974)

Family Educational and Privacy Act (1978)

Privacy Protection Act (1980)

Cable Communication Policy Act (1984)

Video Privacy Protection Act (1988)

Aviation and Transportation Security Act (2001)

Uniting and Strengthening America by Providing Appropriate Tools Required to Intercept and Obstruct Terrorism Act (USA Patriot Act) (2001)

Enhanced Border Security and Visa Entry Reform Act (2002)

Real ID Act (2005)

Biometrics Enhancement and Airport-Risk Reduction Act (BEAR) Act (2008)

European

Council of Europe (1950–1998) *Convention for the Protection of Human Rights and Fundamental Freedoms*

Council of Europe (1981) *Convention for the Protection of Individuals with Regard to Automatic Processing of Personal Data*

Directive 95/46/EC of the European Parliament and of the Council of 24 October 1995 on the protection of individuals with regard to the processing of personal data and on the free movement of such data

Council of Europe (1997) *Convention for the Protection of Human Rights and Dignity of the Human Being with regard to the Application of Biology and Medicine: Convention on Human Rights and Biomedicine*, (Oviedo)

Charter of Fundamental Rights of the European Union (2000)

Council of Europe, *Council Decision of 29 May 2000 concerning the request of the United Kingdom of Great Britain and Northern Ireland to take part in some of the provisions of the Schengen acquis (2000/365/EC)*

Council Regulation (EC) No 2725/2000 of 11 December 2000 concerning the establishment of 'Eurodac' for the comparison of fingerprints for the effective application of the Dublin Convention

Regulation (EC) No 45/2001 of the European Parliament and the Council of 18 December 2000 *on the protection of individuals with regard to the processing of personal data by the Community institutions and bodies and on the free movement of such data*

Council of Europe, *Council Decision of 28 February 2002 concerning Ireland's request to take part in some of the provisions of the Schengen acquis (2002/192/EC)*

Council of Europe, *(EC) No 2252/2004 of 13 December 2004 on standards for security features and biometrics in passports and travel documents issued by Member States*

Council of Europe, (EC) No 444/2009 *of 28 May 2009 amending Council Regulation (EC) No 2252/2004 on standards for security features and biometrics in passports and travel documents issued by Member States*

International

Charter of the United Nations (1945)

United Nations, *Universal Declaration of Human Rights (1948)*

United Nations, *International Covenant on Civil and Political Rights (1976)*

Legal Cases

Dagg v. *Canada (Minister of Finance)* [1997] 2 S.C.R. 403

Haughey v. *Moriarty* [1999] 3 I.R. 1

Katz v. *United States*, 386 U.S. 954 (1967)

Kennedy and Arnold v. *Ireland* [1987] I.R. 587: [1988] ILRM 472

McGee v. *Attorney General [1974] IR 284*

S. and Marper v. *United Kingdom* [2008] ECHR 30562/04 [Grand Chamber]

Sanchez v. *The State of New York*, #2002-001-034, Motion No. M-64552

Glossary

Note that the terms listed are explained as they apply in the context of the present document. In broader, more general use, some of the terms will have a wider meaning.[989]

Acceptability: An individual's willingness to accept the use of a particular biometric modality for the purposes of biometric recognition.

Algorithm: A limited sequence of instructions or steps that tells a computer system how to solve a particular problem. A biometric system will have a number of different algorithms for different functions, for example, template generation and matching.

Authentication: See Verification.

Behavioural Biometric: A biometric modality that is learned or acquired over time rather than one based predominantly on an individual's biology (*e.g.* keystroke dynamics or signature).

Biometric Characteristic: See Biometric Modality

Biometric Modality: Any measurable, physical or physiological feature or behavioural trait that can be used to identify an individual or to verify the claimed identity of an individual.

Centralised Storage: Where the biometric information, whether in the form of raw images or templates, from all those individuals enrolled in a particular biometric application is stored in a database.

Collectability: The degree of difficulty associated with presenting and measuring a particular biometric modality quantitatively.

Covert Collection: Where an individual's biometric information is collected or captured without his/her knowledge or express consent.

Cryptosystem: A system in which information is encrypted using a particular key (code) and another PIN or password is required to unlock this key, which can then be used to decode the encrypted information. In a biometric cryptosystem, a biometric template as opposed to a PIN or password is used to unlock the encryption key.

Data Mining: Data research and analysis aiming to extract hidden trends or correlations from large data sets or to identify strategic information.

Distinctiveness: The degree to which a biometric modality differs between different individuals within the population.

989 Several definitions were derived from http://www.biometrics.gov/Documents/Glossary.pdf, and from European Commission Joint Research Centre, Institute for Prospective Technology Studies (2005). *Biometrics at the Frontiers: Assessing the Impact on Society.* Seville, 166p.

Encryption: Transforming information into an unintelligible format so that it cannot be read by unauthorised individuals. A key (code) or password is required to decode the encrypted information.

Enrolment: The initial process of collecting a sample of biometric information from an individual, and storing this information as a digital reference image or template for future comparison when he/she next uses the system.

Failure to Acquire: The failure of a biometric system to capture and/or extract usable information from a biometric sample for comparison with a previously enrolled image or template.

Failure to Enrol: The failure of a biometric system to form a proper enrolment reference (*i.e.* raw image or template) for an individual. This could occur because a biometric sample could not be collected or due to problems with extracting sufficient features from a sample to create a template.

Fallback Procedures: Mechanisms put in place to deal with any problems or errors (*e.g.* failure to acquire errors or false rejection errors) that arise with a particular biometric system.

False Accept Rate: The rate at which a biometric sample collected from one individual, who is not in the system, is mistakenly matched to an image or template from another individual who is enrolled in the system.

False Match Rate: See False Accept Rate.

False Non-Match Rate: See False Reject Rate.

False Reject Rate: The rate at which a biometric sample collected from one individual does not match the enrolled image or template for that individual in the system.

Feature Extraction: Where the biometric system extracts the distinctive discriminatory characteristics (features) from an individual's biometric sample and then uses these characteristics to generate a reference template for that individual.

Fingerprint Minutiae: Discontinuities that disrupt the flow of fingerprint ridges, *e.g.* a ridge ending or a bifurcation.

Function Creep: Where information originally collected for one particular purpose is subsequently used for another previously non-specified purpose.

Identification: This is where a biometric system attempts to ascertain who an individual is without that individual claiming a particular identity. In this case the sample biometric is compared with all the raw images or templates in a given database.

Identity Theft: A criminal activity whereby someone obtains another individual's personal information without that individual's consent or knowledge, and uses this information to impersonate that individual in order to conduct fraudulent transactions.

Interoperability: The degree to which templates generated using a specific algorithm from one biometric system can be used in another unrelated biometric system.

Intra-class Variation: The way in which two samples of the same biometric modality from the same person are never absolutely identical.

Liveness Detection: A method of checking if the biometric sample is being read from a live person as opposed to a fake body part or the body part from a dead person. Liveness detection can involve checking different physiological signs such as blood pressure, pulse rate, respiration, skin conductivity and temperature.

Localised Storage: Where the biometric information, whether in the form of raw images or templates, of an individual enrolled in a biometric system is stored on a portable medium (*e.g.* a smart card or biometric passport), which he/she retains possession of, as opposed to being stored in a database. This is also referred to as decentralised storage.

Logical Access: The process of accessing or logging on to a computer, network or database.

Matching: The process of comparing a biometric sample against a previously stored raw image or template and scoring the level of similarity. If this score is above a predetermined threshold, the newly submitted sample is deemed to match the stored raw image or template.

Match-on-Card System: A system where the biometric information (*i.e.* raw image or template), and the feature extraction and matching modules are all stored and conducted on a smart card. Therefore, the biometric information never leaves the card.

Multimodal Biometric System: A biometric system which uses two or more biometric modalities (*e.g.* face and fingerprint) from the same individual in the recognition process.

One-to-Many Comparison: See Identification.

One-to-One Comparison: See Verification.

Performance: The level of accuracy and speed of recognition of a biometric system, given the operational and environmental factors involved.

Permanence: The degree to which a given biometric modality remains unchanged throughout an individual's life.

Physical/Physiological Biometric: A biometric modality that is based primarily on an anatomical or physiological characteristic of the body, *e.g.* fingerprint or iris.

Raw Image: The raw information collected from an individual by the sensor of a biometric system, *e.g.* an image of a fingerprint or face. This information is stored in a digital format.

Recognition: A generic term used in the description of biometric systems (*e.g.* facial recognition or iris recognition), which relates to their fundamental function. The term recognition can be applied to systems based on identification or verification.

Resistance to Circumvention: The degree of difficulty required to defeat or bypass a particular biometric system.

Screening: A form of biometric identification where an individual's biometric sample is compared with a specific watch list.

Smart Card: A card shaped portable data carrying device, which contains a microchip that can be used to both store and process data.

Spoofing: Where an adversary (illegitimate user) intentionally attempts to fool a biometric system into recognising him/her as a legitimate user of the system, *e.g.* through the use of fake fingers or fingerprints.

Template: Digital data representing the distinct features extracted from an individual's biometric sample. In basic terms, templates take the form of numeric data.

Threshold: A preset value used to determine the margin of error tolerated by the matching algorithm of a biometric system. For the system to declare a match, the matching score needs to be above the designated threshold. Decreasing the threshold makes a system more tolerant to user variations, whereas increasing it makes a system more secure.

Unimodal Biometrics: A biometric system which uses a single biometric modality from an individual during the recognition process.

Universality: The degree to which the population possesses a given biometric modality.

Verification: This is where the biometric system authenticates an individual's claimed identity by comparing the newly collected sample biometric data with the corresponding enrolled template.

Bibliography

Acharya L (2006). *Biometrics and Government.* The Parliamentary Information and Research Service, Library of Parliament, Canada, 19p.

Action on Rights for Children (ARCH) (2007). *Child Tracking: Biometrics in Schools & Mobile Location Devices.* Available online at: http://www.arch-ed.org/issues/Tracking%20devices/final_report_on_child_tracking.htm, accessed 16 May 2008.

Adler A and Schuckers ME (2007). Comparing Human and Automatic Face Recognition Performance. *IEEE Transactions on Systems, Man, and Cybernetics – Part B: Cybernetics* 37(5): 1248–1255.

Albrecht A, Behrens M, Mansfield T, McMeehan W, Rejman-Greene M, Savastano M, Statham P, Schmidt C, Schouten B and Walsh M (2003). *BIOVISION: Roadmap for Biometrics in Europe 2010.* Final Report of the Roadmap Task, D2.6/Issue 1.1. Available online at: http://ftp.cwi.nl/CWIreports/PNA/PNA-E0303.pdf, accessed 9 July 2008.

Alterman A (2003). "A piece of yourself": Ethical issues in biometric identification. *Ethics and Information Technology* 5(3): 139–150.

Agamben G (2004). No to Bio-Political Tattooing. *Le Monde* 10 January 2004.

Antonelli A, Cappelli R, Maio D and Maltoni D (2006). Fake Finger Detection by Skin Distortion Analysis. *IEEE Transactions on Information Forensics and Security* 1(3): 360–373.

Article 29 Data Protection Working Party (2003). *Working Document on Biometrics.* European Commission, Brussels, 11p. Available online at: http://ec.europa.eu/justice_home/fsj/privacy/docs/wpdocs/2003/wp80_en.pdf, accessed 1 November 2007.

Article 29 Data Protection Working Party (2005). *Opinion on Implementing the Council Regulation (EC) No 2252/2004 of 13 December 2004 on standards for security features and biometrics in passports and travel documents issued by Member States.* European Commission, Brussels, 12p. Available online at: http://ec.europa.eu/justice_home/fsj/privacy/docs/wpdocs/2005/wp112_en.pdf, accessed 17 November 2008.

Australian Government, Department of Immigration and Citizenship (2007). *Identity Matters: Strategic Plan for Identity Management in DIAC 2007–2010.* Department of Immigration and Citizenship, Canberra, 65p.

Australian Government, Department of Immigration and Citizenship (2008). *Fact Sheet 2. Key Facts in Immigration.* Available online at: http://www.immi.gov.au/media/fact-sheets/02key.htm, accessed 14 November 2008.

Australian Government, Department of Immigration and Citizenship (2008). *Fact Sheet 84 – Biometric Initiatives.* Available online at: http://www.immi.gov.au/media/fact-sheets/84biometric.htm, accessed 14 November 2008.

Australian Government, Office of the Privacy Commissioner (2007). *The Operation of the Privacy Act Annual Report 1 July 2006 – 30 June 2007.* Office of the Privacy Commissioner, Sydney, 167p. Available online at: http://www.privacy.gov.au/publications/07annrep/07annrep.pdf, accessed 14 November 2008.

Australian Law Reform Commission (2008). *For Your Information: Australian Privacy Law and Practice,* Volume 1, Report 108. Australian Law Reform Commission, Sydney, 830p. Available online at: http://www.austlii.edu.au/au/other/alrc/publications/reports/108/vol_1_full.pdf, accessed 17 November 2008.

Australian Law Reform Commission (2008). *Technology-neutral privacy principles should govern rapidly developing ICT.* Press release from the Australian Law Reform Commission published 11 August 2008. Available online at: http://www.alrc.gov.au/media/2008/mbn2.pdf, accessed 17 November 2008.

BBC News (2004). *Barcelona clubbers get chipped.* BBC News, published online 29 September 2004. Available online at: http://news.bbc.co.uk/2/hi/technology/3697940.stm, accessed 8 April 2008.

Becta (2007). *Becta guidance on biometric technologies in schools.* Becta, Coventry, 10p. Available online at: http://schools.becta.org.uk/upload-dir/downloads/becta_guidance_on_biometric_technologies_in_schools.doc, accessed 5 November 2007.

Berlin I (1969). *Four Essays on Liberty.* Oxford University Press, Oxford, 32p. Available online at: http://www.nyu.edu/projects/nissenbaum/papers/twoconcepts.pdf, accessed 20 November 2008.

Biever C (2005). ID revolution – prepare to meet the new you. *New Scientist* 187(2516): 26–29.

Bigo D, Carrera S, Guild E and Walker RBJ (2007). *The Changing Landscape of European Liberty and Security: Mid-Term Report on the Results of the CHALLENGE Project.* Research Paper No. 4, 46p. Available online at: http://www.libertysecurity.org/article1357.html, accessed 9 July 2008.

Biometric Information Technology Ethics (2005). *Biometrics and Privacy.* Report of the Second BITE Scientific Meeting, Tuesday 26th April 2005, Rome, Italy, 13p. Available online at: http://www.biteproject.org/documents/report_biometrics_privacy.pdf, accessed 16 October 2007.

Biometric Technology Today (2006). Banco Azteca rolls out biometrics to 8m customers. *Biometric Technology Today* 14(5): 4.

Biometric Technology Today (2007a). Face Recognition: Part 1. *Biometric Technology Today* 15(9): 11.

Biometric Technology Today (2007b). Face Recognition: Part 2. *Biometric Technology Today* 15(10): 10–11.

Bowyer KW (2004). Face Recognition Technology: Security versus Privacy. *IEEE Technology and Society Magazine* Spring 2004: 9–20.

Bowyer KW, Chang KI, Flynn PJ and Chen X (2006). Face Recognition Using 2-D, 3-D, and Infrared: Is Multimodal Better than Multisample? *Proceedings of the IEEE* 94(11): 2000–2012.

Bowyer KW, Chang KI, Yan P, Flynn PJ, Hansley E and Sarkar S (2006). *Multi-Modal Biometrics: An Overview*. Second Workshop on Multi-Modal User Authentication (MMUA 2006), May 2006, Toulouse, France, 8p. Available online at: http://www.nd.edu/~kwb/BowyerEtAlMMUA_2006.pdf, accessed 22 February 2008.

Burge M and Burger W (1999). Ear Biometrics. In A Jain, R Bolle and S Pankanti (eds.) *Biometrics: Personal Identification in Networked Society*, Kluwer Press, Dordrecht, p.273–286.

Campbell JP (1999). Speaker Recognition. In A Jain, R Bolle and S Pankanti (eds.) *Biometrics: Personal Identification in Networked Society*, Kluwer Press, Dordrecht, p.165–190.

Cavoukian A (1999). *Privacy and Biometrics*. Information and Privacy Commissioner/Ontario, Toronto, 15p. Available online at: http://www.ipc.on.ca/images/Resources/pri-biom.pdf, accessed 6 February 2008.

Cavoukian A (2005). *Identity Theft Revisited: Security is Not Enough*. Information and Privacy Commissioner/Ontario, Toronto, 39p. Available online at: http://www.ipc.on.ca/images/Resources/idtheft-revisit.pdf, accessed 7 February 2008.

Cavoukian A (2006). *7 Laws of Identity: The Case for Privacy-Embedded Laws of Identity in the Digital Age*. Information and Privacy Commissioner/Ontario, Toronto, 18p. Available online at: http://www.ipc.on.ca/images/Resources/up-7laws_whitepaper.pdf, accessed 7 February 2008.

Cavoukian A and Stoianov A (2007). *Biometric Encryption: A Positive-Sum Technology that Achieves Strong Authentication, Security AND Privacy*. Information and Privacy Commissioner/Ontario, Toronto, 48p. Available online at: http://www.ipc.on.ca/images/Resources/bio-encryp.pdf, accessed 6 February 2008.

Chan ADC, Hamdy MM, Badre A and Badee V (2006). Person Identification Using Electrocardiograms. *Canadian Conference on Electrical and Computer Engineering 2006 (CCECE '06)* May 2006: 1–4.

Cherry S (2007). Personal biometrics, private data. *Password, Philips Research Technology Magazine* 30: 5–8.

The Chubb Corporation (2005). *One in Five Americans Has Been a Victim of Identity Fraud.* Available online at: http://www.chubb.com/corporate/chubb3875.html, accessed 12 January 2009.

Clarke DM (1984). *Church and State: Essays in Political Philosophy.* Cork University Press, Cork, 275p.

Collins J (2007). Data insecurities. *The Irish Times* 23 November 2007. Available online at: http://www.irishtimes.com/newspaper/finance/2007/1123/1195682113657.html, accessed 4 February 2008.

Commission de l'éthique de la science et de la technologie (2008). *In Search of Balance: An Ethical Look at New Surveillance and Monitoring Technologies for Security Purposes.* Commission de l'éthique de la science et de la technologie, Quebec, 73p.

Commission of the European Communities, Commission Staff Working Document (2004). *The implementation of Commission Decision 520/2000/EC on the adequate protection of personal data provided by the Safe Harbour privacy Principles and related Frequently Asked Questions issued by the US Department of Commerce*, SEC (2004) 1323. Commission of the European Communities, Brussels, 14p. Available online at: http://ec.europa.eu/justice_home/fsj/privacy/docs/adequacy/sec-2004-1323_en.pdf, accessed 5 November 2008.

Council of the European Union (2004). *The Hague Programme: strengthening freedom, security and justice in the European Union*, 16054/04. Council of the European Union, Brussels, 33p. Available online at:
http://ec.europa.eu/justice_home/doc_centre/doc/hague_programme_en.pdf, accessed 17 November 2008.

Coyle C (2008). Irish bus pass is 'identity card by stealth'. *The Sunday Times* 10 August 2008. Available online at: http://www.timesonline.co.uk/tol/news/world/ireland/article4493788.ece, accessed 14 August 2008

Crews CW Jr (2002). Human Bar Code: Monitoring Biometric Technologies in a Free Society. *Policy Analysis* 452: 1–20. Available online at: http://www.cato.org/pubs/pas/pa452.pdf, accessed 30 May 2008.

Crowley MG (2006). *Cyber crime and biometric authentication – the problem of privacy versus the protection of business assets.* 8p. Available online at: http://igneous.scis.ecu.edu.au/proceedings/2006/aism/Crowley%20-%20Cyber%20crime%20and%20biometric%20authentication%20the%20problem%20of%20privacy%20versus%20protection%20of%20business%20assets.pdf, accessed 21 November 2008.

Data Protection Commissioner (2007). *Biometrics in Schools, Colleges and other Educational Institutions.* Available online at: http://www.dataprotection.ie/docs/Biometrics_in_Schools_Colleges_and_other_Educational_Institu/409.htm, accessed 5 November 2007.

Data Protection Commissioner (2008a). *Annual Report of the Data Protection Commissioner 2007.* Brunswick Press Ltd, Dublin, 88p. Available online at: http://www.dataprotection.ie/documents/annualreports/AR2007En.pdf, accessed 13 May 2008.

Data Protection Commissioner (2008b). *Data Protection in the Department of Social & Family Affairs.* Report by the Data Protection Commissioner. Data Protection Commissioner, Portarlington, 37p. Available online at: http://www.welfare.ie/EN/Topics/Documents/ODPCReport.pdf, accessed 1 August 2008.

Data Protection Commissioner (2008c). *Biometrics in the workplace.* Available online at: http://www.dataprotection.ie/docs/Biometrics_in_the_workplace./244.htm, accessed 13 May 2008.

Data Protection Commissioner (2009). *Twentieth Annual Report of the Data Protection Commissioner 2008.* Data Protection Commissioner, Portarlington, 111p. Available online at: http://www.dataprotection.ie/documents/annualreports/AR2008.pdf, accessed 14 May 2008.

Daugman J (2004). How Iris Recognition Works. *IEEE Transactions on Circuits and Systems for Video Technology* 14(1): 21–30.

Daugman J (2008). *United Arab Emirates Deployment of Iris Recognition.* Available online at: http://www.cl.cam.ac.uk/~jgd1000/deployments.html, accessed 15 February 2008.

Daugman J and Malhas I (2004). Iris recognition border-crossing system in the UAE. *International Airport Review* 2: 49–53.

Davies SG (1994). Touching Big Brother: How biometric technology will fuse flesh and machine. *Information Technology & People* 7(4): 38–47.

Davies S, Hosein I and Whitley EA (2005). *The Identity Project: An assessment of the UK Identity Cards Bill and its implications.* The London School of Economics and Political Science London, 303p. Available online at: http://eprints.lse.ac.uk/684/1/identityreport.pdf, accessed 15 February 2008.

de Bréadún D (2008). Ahern sets up group to review data protection law. *The Irish Times* 1 November 2008. Available online at: http://www.irishtimes.com/newspaper/ireland/2008/1101/1225321623499_pf.html, accessed 14 November 2008.

De Hert P, Schreurs W and Brouwer E (2007). Machine readable identity documents with biometric data in the EU – part III. *Keesing Journal of Documents & Identity* 23: 27–32.

Delany H (2008). *The Right to Privacy. A Doctrinal and Comparative Analysis.* Thomson Round Hall, Dublin, 352p.

Department of Homeland Security (2004). *US-VISIT Program Privacy Policy.* Department of Homeland Security, Washington, 5p. Available online at: http://www.dhs.gov/xlibrary/assets/privacy/privacy_stmt_usvisit.pdf, accessed 11 April 2008.

Department of Homeland Security (2007). *Privacy Impact Assessment Update for the Conversion to 10-Fingerprint Collection for the United States Visitor and Immigration Status Indicator Technology Program (US-VISIT).* Department of Homeland Security, Washington, 15p. Available online at: http://www.dhs.gov/xlibrary/assets/privacy/privacy_pia_usvisit_10p.pdf, accessed 22 May 2008.

Department of Homeland Security (2008). *DHS Releases REAL ID Regulation.* Press release from the Department of Homeland Security 11 January 2008. Available online at: http://www.dhs.gov/xnews/releases/pr_1200065427422.shtm, accessed 3 November 2008.

Department of Social and Family Affairs (2008). *Personal Public Service Number.* SW 100, Department of Social and Family Affairs, Sligo, 4p. Available online at: http://www.welfare.ie/EN/Publications/SW100/Documents/sw100.pdf, accessed 14 August 2008.

Dessimoz D, Richiardi J, Champod C and Drygajlo A (2006). *Multimodal Biometrics for Identity Documents (MBioID): State-of-the-Art* (Version 2.0), Research Report, PFS 341-08.05, Institut de Police Scientifique – Ecole des Sciences Criminelles (Université de Lausanne) & Speech Processing and Biometric Group (Ecole Polytechnique Fédérale de Lausanne), 156p. Available online at: http://www.europeanbiometrics.info/images/resources/90_264_file.pdf, accessed 7 February 2008.

Dhont J, Asinari MVP and Poullet Y (April 2004). *Safe Harbour Decision Implementation Study,* at the request of the European Commission, Internal Market DG. Available online at: http://ec.europa.eu/justice_home/fsj/privacy/docs/studies/safe-harbour-2004_en.pdf, accessed 5 November 2008

Digital Persona, Inc. (2006). *Digital Persona Deploys World's Largest Biometric Banking Application.* Available online at: http://www.digitalpersona.com/index.php?id=pr_20060322, accessed 12 January 2009.

Dike-Anyiam B and Rehmani Q (2006). Biometric vs. Password Authentication: A User's Perspective. *The Journal of Information Warfare* 5(1): 33–45.

Donnolly M (2002). *Consent: Bridging the Gap between Doctor and Patient.* Cork University Press, Cork, 96p.

The Economist (2003). Biometrics: Too flaky to trust. *The Economist* 4 December 2003. Available online at:

http://www.economist.com/opinion/displaystory.cfm?story_id=E1_NNGGNJD, accessed 17 October 2007.

The Economist (2005). Canada and the United States: The Unfriendly Border. *The Economist* 376(8441): 31–32.

The Economist (2006). Biometrics gets down to business. *The Economist* 30 November 2006. Available online at:
http://www.economist.com/science/tq/displaystory.cfm?story_id=E1_RPTNNQG, accessed 17 October 2007.

The Economist (2007). Mobile commerce. A better way of paying by phone. *The Economist* 19 July 2007. Available online at:
http://www.economist.com/science/PrinterFriendly.cfm?story_id=9507446, accessed 17 October 2007.

Electronic Privacy Information Center and Privacy International (2007). *Privacy and Human Rights 2006. An International Survey of Privacy Laws and Developments.* Electronic Privacy Information Center and Privacy International, US. Available online at:
http://www.privacyinternational.org/article.shtml?cmd[347]=x-347-559458, accessed 22 September 2008.

Etzioni A (1999). *The Limits of Privacy.* Basic Books, New York, 288p.

Eurodac Supervision Coordination Group (2007). *EURODAC Supervision Coordination Group, Report of the first coordinated inspection.* European Data Protection Supervisor, Brussels, 16p.

European Commission Joint Research Centre, Institute for Prospective Technology Studies (2005). *Biometrics at the Frontiers: Assessing the Impact on Society.* Seville, 166p.

European Data Protection Supervisor (2008). Opinion of the European Data Protection Supervisor on the proposal for a Regulation of the European Parliament and of the Council amending Council Regulation (EC) No 2252/2004 on standards for security features and biometrics in passports and travel documents issued by Member States, (2008/C 200/01). *Official Journal of the European Union* C200: 1–5. Available online at: http://www.edps.europa.eu/EDPSWEB/webdav/site/mySite/shared/Documents/Consultation/Opinions/2008/08-03-26_Biometrics_passports_EN.pdf, accessed 17 November 2008.

European Digital Rights (2009). *Lucky Win For The Swiss Biometric Passports.* Available online at: http://www.edri.org/edri-gram/number7.20/swiss-biometric-passports, accessed 1 July 2009.

Fried C (1984). Privacy [A moral analysis]. In FD Schoeman (ed.) Philosophical *Dimensions of Privacy. An Anthology.* Cambridge University Press, New York, p.203–222.

Futurelab (2007). Should we allow Big Brother in schools? *Vision* 4: 1–4. Available online at: http://www.futurelab.org.uk/resources/documents/vision/VISION_04.pdf, accessed 16 May 2008.

Geiser U (2009a). *Swiss vote on introduction of biometric passports.* Available online at: http://www.swissinfo.ch/eng/front/Swiss_vote_on_introduction_of_biometric_passports.html?siteSect=105&sid=10707459&rss=true&ty=st, accessed 1 July 2009.

Geiser U (2009b). Passport Vote Wins Majority and Puzzles Experts. *The Journal of Turkish Weekly* 17 May 2009. Available online at: http://www.turkishweekly.net/news/77082/-passport-vote-wins-majority-and-puzzles-experts.html, accessed 1 July 2009.

Graham-Rowe D (2005). Privacy and prejudice: whose ID is it anyway? *New Scientist* 187(2517): 20–23.

Gutwirth S (2007). Biometrics between opacity and transparency. *Annali dell Institute Superiore di Sanitá* 43(1): 61–65.

Harden B (2004). FBI Faulted in Arrest of Ore. Lawyer. *The Washington Post* 16 November 2004. Available online at: http://www.washingtonpost.com/wp-dyn/articles/A52907-2004Nov15.html, accessed 12 January 2009.

Harel A (2009). Biometrics, Identification and Practical Ethics. In E Mordini and M Green (eds.) *Identity, Security and Democracy: The Wider Social and Ethical Implications of Automated Systems for Human Recognition.* Volume 49 NATO Science for Peace and Security Series – E: Human and Societal Dynamics, IOS Press, Amsterdam, p.69–84.

Henderson SC and Snyder CA (1999). Personal information privacy: implications for MIS managers. *Information & Management* 36(4): 213–220.

Heyer R (2008). *Biometric Technology Review 2008.* Land Operations Division, (DSTO) Defence Science and Technology Organisation, Australia, 60p.

Henderson SC and Snyder CA (1999). Personal information privacy: implications for MIS managers, *Information and Management*, 24(4): 213–220.

Henschke A (2007). *An Evaluation of Forensic DNA Databases Using Different Conceptions of Identity.* MSc Thesis, Linköping University, Sweden, 71p. Available online at: http://liu.diva-portal.org/smash/record.jsf?pid=diva2:23785, accessed 23 July 2008.

Hill R (1999). Retina Identification. In A Jain, R Bolle and S Pankanti (eds.) *Biometrics: Personal Identification in Networked Society*, Kluwer Press, Dordrecht, p.123–142.

Hodges S and McFarlane D (2005). *Radio frequency identification: technology, applications and impact.* Auto-ID Labs White Paper Series, Edition 1. Available online at: http://www.autoidlabs.org/single-view/dir/article/6/60/page.html, accessed 8 April 2008.

Hong L and Jain AK (1999). Multimodal Biometrics. In A Jain, R Bolle and S Pankanti (eds.) *Biometrics: Personal Identification in Networked Society*, Kluwer Press, Dordrecht, p.327–344.

House of Commons Home Affairs Committee (2008). *A Surveillance Society? HC-I, Fifth Report of Session 2007–08. Volume I: Report, together with formal minutes*. The Stationery Office Limited, London, 117p.

Howells L (2005). *Fusion Comes in from the Cold.* A Consult Hyperion White Paper. Consult Hyperion, Surrey, 15p. Available online at: http://www.chyp.com/PubWebFiles/whitepaper/fusion_comes_in_from_cold.pdf, accessed 7 February 2008.

Huijgens R (2006). Technology trends – 2006. In Schmitz PE, Tavano R, Lodge J, Huijgens R, Aisola K and Flammang M (eds.). *Biometrics in Europe. Trend Report 2006*. UNISYS, Brussels, 113p.

Huijgens R (2007). Trends in biometrics technology. In Schmitz PE, Huijgens R and Flammang M (eds.). *Biometrics in Europe. Trend Report* 2007. UNISYS, Brussels, 39p.

Hunter S (2003). A critical analysis of approaches to the concept of social identity in social policy. *Critical Social Policy* 23(3): 322–344.

Iachello G and Abowd GD (2005). Privacy and Proportionality: Adapting Legal Evaluation Techniques to Inform Design in Ubiquitous Computing. *CHI* 2005, April 2–7 2005, Portland Oregon, USA, p.91–100. Available online at: http://luci.ics.uci.edu/predeployment/websiteContent/weAreLuci/biographies/faculty/djp3/LocalCopy/p91-iachello.pdf, accessed 21 November 2008.

Information Commissioner's Office (2008). *The use of biometrics in schools.* V1.1 August 2008. Information Commissioner's Office, Cheshire, UK, 3p. Available online at: http://www.ico.gov.uk/upload/documents/library/data_protection/detailed_specialist_guides/fingerprinting_final_view_v1.11.pdf, accessed 15 September 2008.

International Biometric Group (2007a). *BioPrivacy Best Practices.* International Biometric Group, New York. Available online at: http://www.biometricgroup.com/reports/public/reports/privacy_best_practices.html, accessed 15 February 2008.

International Biometrics Group (2007b). *Generating Images from Templates.* International Biometric Group, New York. Available online at: http://www.biometricgroup.com/reports/public/reports/templates_images.html, accessed 15 February 2008.

International Biometrics Group (2007c). *Liveness Detection in Biometric Systems.* International Biometric Group, New York. Available online at: http://www.biometricgroup.com/reports/public/reports/liveness.html, accessed 15 February 2008.

International Biometric Group (2007d). *Fingerprint Feature Extraction.* International Biometric Group, New York. Available online at:
http://www.biometricgroup.com/reports/public/reports/finger-scan_extraction.html, accessed 18 April 2008.

International Biometric Group (2007e). *Optical – Silicon – Ultrasound.* International Biometric Group, New York. Available online at:
http://www.biometricgroup.com/reports/public/reports/finger-scan_optsilult.html, accessed 6 March 2008.

International Civil Aviation Organization Technical Advisory Group (ICAO TAG) (2004). *Biometrics Deployment of Machine Readable Travel Documents.* International Civil Aviation Organization Technical Advisory Group MRTD/NTWG, Technical Report, Version 2, 60p.

International Commission of Jurists (2009). *Assessing Damage, Urging Action. Report of the Eminent Jurists Panel on Terrorism, Counter-terrorism and Human Rights.* International Commission of Jurists, Geneva, 199p.

Irish Council for Bioethics (2007). *Is It Time For Advance Healthcare Directives? Opinion.* Irish Council for Bioethics, Dublin, 98p. Available online at:
http://www.bioethics.ie/uploads/docs/Advance_Directives_HighRes.pdf.

Irish Council for Civil Liberties (2008). *Safeguards Essential in Piloting of Public Service Card Says the ICCL.* Press Release from the Irish Council for Civil Liberties, 5 August 2008.

Irish Human Rights Commission (2008). *Observations on the Immigration, Residence and Protection Bill 2008.* Irish Human Rights Commission, Dublin, 126p.

The Irish Times (2007). Surveillance Nation. *The Irish Times* 4 August 2007. Available online at:
http://www.irishtimes.com/newspaper/newsfeatures/2007/0804/1186123297026.html, accessed 9 November 2007.

Jain A, Bolle R and Pankanti S (1999). Introduction to Biometrics. In A Jain, R Bolle and S Pankanti (eds.) *Biometrics: Personal Identification in Networked Society*, Kluwer Press, Dordrecht, p.1–42.

Jain AK, Nandakumar K and Nagar A (2008). Biometric Template Security. *EURASIP Journal on Advances in Signal Processing. Special Issue Advanced Signal Processing and Pattern Recognition Methods for Biometrics* Volume 2008, Article ID 579416, 17p.

Jain AK, Ross A and Pankanti S (2006). Biometrics a Tool for Information Security. *IEEE Transactions on information Forensics and Security* 1(2): 125–143.

Jain AK, Ross A and Prabhakar S (2004). An Introduction to Biometric Recognition. *IEEE Transactions on Circuits and Systems for Video Technology* 14(1): 4–20.

Jain AK, Ross A and Uludag U (2005). Biometric Template Security: Challenges and Solutions. *Proceedings of the European Signal Processing Conference (EUSIPCO '05), Antalya, Turkey, September 2005*, 4p. Available online at: http://biometrics.cse.msu.edu/Publications/SecureBiometrics/JainRossUludag_TemplateSecurity_EUSIPCO05.pdf, accessed 31 March 2008.

Karp J and Meckler L (2006). Which Travelers Have 'Hostile Intent'? Biometric Device May Have the Answer. *The Wall Street Journal* 14 August 2006. Available online at: http://online.wsj.com/public/article/SB115551793796934752-2hgveyRtDDtssKozVPmg6RAAa_w_20070813.html?mod=tff_main_tff_top, accessed 5 December 2008.

King P (1996). *Socialism and the Common Good. New Fabian Essays*. Taylor and Francis Inc., London, 336p.

Kurniawan SH (2007). *Bringing convenience and security to everyday life – Case Histories: A1 Team Malaysia and Elderly Home Association*. Presentation at the Biometrics Exhibition and Conference 2007, 17–19 October 2007, Westminster, London.

Law Reform Commission (2005). *Report – The Establishment of a DNA Database*. (LRC 78-2005), Law Reform Commission, Dublin, 129p.

Legomsky SH (2005). The ethnic and religious profiling of noncitizens: national security and human rights. *Boston College Third World Law Journal* 25(1): 161–196.

Lodge J (2007). Freedom, security and justice: the thin end of the wedge for biometrics? *Annali dell Institute Superiore di Sanitá* 43(1): 20–26.

Lowe AL, Urquhart A, Foreman LA, and Evett IW (2001). Inferring ethnic origin by means of an STR profile. *Forensic Science International* 119(1): 17–22.

Lyon D (2008). Biometrics, Identification and Surveillance. *Bioethics* 22(9): 499–508.

MacDonald F (2008). A card up Africa's sleeve. *Metro* 26 February 2008.

MacIntyre A (1994). The Concept of a Tradition. In M Daly (ed.) *Communitarianism: A New Public Ethics*. Wadsworth Publishing Company, California, p.123–126.

Mahony H (2007). Database of passenger flight details proposed. *The Irish Times* 7 November 2007. Available online at: http://www.irishtimes.com/newspaper/world/2007/1107/1194222776758.html, accessed 8 November 2007.

Marks P (2007). Can a government remotely detect terrorists thoughts? *New Scientist* 195(2616): 24–25.

Marshall P (2007). We can see clearly now. *Government Computer News* June 4, 2007. Available online at: http://www.itl.nist.gov/iad/News/FaceRecog3.html, accessed 23 May 2008

Martinez E (2007). *Case History: Spanish police implement world's first automatic speaker identification system (ASIS).* Presentation at the Biometrics Exhibition and Conference 2007, 17–19 October 2007, Westminster, London, and Agnitio website

Matey JR, Naroditsky O, Hanna K, Kolczynski R, LoIacono DJ, Mangru S, Tinker M, Zappia TM and Zhao WY (2006). Iris on the Move: Acquisition of Images for Iris Recognition in Less Constrained Environments. *Proceedings of the IEEE* 94(11): 1936–1947.

Matsumoto T, Matsumoto H, Yamada K and Hoshino S (2002). Impact of Artificial "Gummy" Fingers on Fingerprint Systems. *Proceedings of SPIE* 4677: 275–289.

McGee H (2008). 16 office laptops stolen since 1999, says comptroller. *The Irish Times* 2 August 2008. Available online at:
http://www.irishtimes.com/newspaper/ireland/2008/0802/1217368897558.html, accessed 2 August 2008.

Michael J (2008). Identity theft crime affects 87,000 – survey. *The Irish Times* published online 6 October 2008. Available online at:
http://www.irishtimes.com/newspaper/breaking/2008/1006/breaking23.htm, accessed 15 October 2008.

Mill JS (1863). *On Liberty*. 2nd Edition. Ticknor and Fields, Boston, 223p.

Mobin UY (2007). *Pakistan's National Biometric Identification Program*. Presentation at the Biometrics Exhibition and Conference 2007, 17–19 October 2007, Westminster, London.

Moor JH (1990). The Ethics of Privacy Protection. *Library Trends* 39(1 and 2): 69–82.

Mordini E (2008). Biometrics, Human Body, and Medicine: A Controversial History. In P. Duquenoy, C. George and K. Kimppa (eds.) *Ethical, Legal, and Social Issues in Medical Informatics*. IGI Global, London, p.249–272.

Mordini E and Massari S (2008). Body, Biometrics and Identity. *Bioethics* 22(9): 488–498.

Mordini E and Ottolini C (2007). Body identification, biometrics and medicine: ethical and social considerations. *Annali dell Istituto Superiore di Sanità* 43(1): 51–60.

Most CM (2003). Battle of the Biometrics. *Digital ID World* November/December 2003: 16–18.

Most CM (2004a). Biometrics and Financial Services – Show Me the Money! *Digital ID World* January/February 2004: 20–23.

Most CM (2004b). Towards Privacy Enhancing Applications of Biometrics. *Digital ID World* June/July 2004: 18–20.

Most CM (2007). *Mega Trends and Meta Drivers for the Biometrics Industry: 2007–2020.* Presentation at the Biometrics Exhibition and Conference 2007, 17–19 October 2007, Westminster, London.

Muller BJ (2004). (Dis)Qualified Bodies: Securitization, Citizenship and 'Identity Management'. *Citizenship Studies* 8(3): 279–294.

Nabeth T and Hildebrandt M (2005). D 2.1: *Inventory of topics and clusters.* Future of Identity in the Information Society (FIDIS) Project Deliverable Version 2.0, 57p. Available online at: http://www.fidis.net/fileadmin/fidis/deliverables/fidis-wp2-del2.1_Inventory_of_topics_and_clusters.pdf, accessed 28 July 2008.

Nalwa VS (1999). Automatic On-line Signature Verification. In A Jain, R Bolle and S Pankanti (eds.) *Biometrics: Personal Identification in Networked Society*, Kluwer Press, Dordrecht, p.143–164

Narveson J (2002). Collective Responsibility. *The Journal of Ethics* 6(2): 179–198.

National Consultative Ethics Committee for Health and Life Sciences (2007). *Biometrics, identifying data and human rights.* Opinion No. 98. National Consultative Ethics Committee for Health and Life Sciences, France, 22p.

National Science and Technology Council (NSTC) Subcommittee on Biometrics (2006a). *Biometrics History.* NSTC, Washington, 27p. Available online at: http://www.biometrics.gov/Documents/BioHistory.pdf, accessed 10 October 2007.

National Science and Technology Council (NSTC) Subcommittee on Biometrics (2006b). *The National Biometrics Challenge.* NSTC, Washington, 19p. Available online at: http://www.biometrics.gov/Documents/biochallengedoc.pdf accessed 10 October 2007.

National Science and Technology Council (NSTC) Subcommittee on Biometrics (2006c). *Biometrics Frequently Asked Questions.* NSTC, Washington, 25p. Available online at: http://www.biometrics.gov/Documents/FAQ.pdf, accessed 10 October 2007.

National Science and Technology Council (NSTC), Subcommittee on Biometrics (2006d). *Privacy & Biometrics: Building a Conceptual Foundation.* NSTC, Washington, 57p. Available online at: http://www.biometrics.gov/docs/privacy.pdf, accessed 10 October 2007.

National Science and Technology Council (NSTC) Subcommittee on Biometrics (2006e). *Biometrics Overview.* NSTC, Washington, 10p. Available online at: http://www.biometricscatalog.org/NSTCSubcommittee/Documents/Biometrics%20Overview.pdf, accessed 10 October 2007.

National Science and Technology Council (NSTC) Subcommittee on Biometrics (2006f). *Palm Print Recognition*. NSTC, Washington, 10p. Available online at: http://www.biometricscatalog.org/NSTCSubcommittee/Documents/Palm%20Print%20Recognition.pdf, accessed 10 October 2007.

National Science and Technology Council (NSTC) Subcommittee on Biometrics (2006g). *Hand Geometry*. NSTC, Washington, 7p. Available online at: http://www.biometrics.gov/Documents/HandGeometry.pdf, accessed 10 October 2007.

National Science and Technology Council (NSTC) Subcommittee on Biometrics (2006h). *Vascular Pattern Recognition*. NSTC, Washington, 6p. Available online at: http://www.biometrics.gov/Documents/VascularPatternRec.pdf, accessed 10 October 2007.

National Science and Technology Council (NSTC) Subcommittee on Biometrics (2006i). *Iris Recognition*. NSTC, Washington, 10p. Available online at: http://www.biometrics.gov/Documents/IrisRec.pdf, accessed 10 October 2007.

National Science and Technology Council (NSTC) Subcommittee on Biometrics (2006j). *Speaker Recognition*. NSTC, Washington, 9p. Available online at: http://www.biometrics.gov/Documents/SpeakerRec.pdf, accessed 10 October 2007.

National Science and Technology Council (NSTC) Subcommittee on Biometrics (2006k). *Dynamic Signature*. NSTC, Washington, 7p. Available online at: http://www.biometrics.gov/Documents/DynamicSig.pdf, accessed 10 October 2007.

National Science and Technology Council (NSTC) Subcommittee on Biometrics (2006l). *International Conference on Biometrics and Ethics, Conference Highlights*. NSTC, Washington, 3p. Available online at: http://www.biometrics.gov/NSTC/Conference_on_Biometrics_and_%20Ethics_2006_highlights.pdf, accessed 28 July 2009.

National Science and Technology Council (NSTC) Subcommittee on Biometrics and Identity Management (2007). *NSTC Policy for Enabling the Development, Adoption and Use of Biometric Standards*. NSTC, Washington, 11p. Available online at: http://www.biometrics.gov/Standards/NSTC_Policy_Bio_Standards.pdf, accessed 2 May 2008.

New Scientist (2005). The myth of fingerprints. *New Scientist* 187(2517): 3.

New Scientist (2009). Encrypted CCTV protects the innocent. *New Scientist* 2717: 19.

Newton EM and Phillips PJ (2007). *Meta-Analysis of Third-Party Evaluations of Iris Recognition*. Technical Report NISTIR 7440, National Institute of Standards and Technology, Maryland, 14p.

Newton G (2004). *DNA Fingerprinting Enters Society*. Available online at: http://genome.wellcome.ac.uk/doc_wtd020878.html, accessed 2 July 2009.

Nixon MS, Carter JN, Cunado D, Huang PS and Stevenage SV (1999). Automatic Gait Recognition. In A Jain, R Bolle and S Pankanti (eds.) *Biometrics: Personal Identification in Networked Society*, Kluwer Press, Dordrecht, p.231–250.

Nuffield Council on Bioethics (2007). *The forensic use of bioinformation: ethical issues.* Nuffield Council on Bioethics, London, 139p. Available online at: http://www.nuffieldbioethics.org/fileLibrary/pdf/The_forensic_use_of_bioinformation_-_ethical_issues.pdf, accessed 31 October 2007.

Nygren S (2007). Non-contact and RFID – not only in logistics. *Detektor International* 3: 14–15.

Obaidat MS and Sadoun B (1999). Keystroke Dynamics Based Authentication. In A Jain, R Bolle and S Pankanti (eds.) *Biometrics: Personal Identification in Networked Society*, Kluwer Press, Dordrecht, p.213–230.

O'Brien C (2008). Photo ID cards may face cash constraint delays. *The Irish Times* 6 August 2008. Available online at: http://www.irishtimes.com/newspaper/ireland/2008/0806/1217923985190.html, accessed 14 August 2008

Office of the Auditor General of Canada (2004). *Report of the Auditor General of Canada to the House of Commons March 2004.* Office of the Auditor General of Canada, Ontario 2004. Available online at: http://www.oag-bvg.gc.ca/internet/docs/20040303ce.pdf, accessed 17 November 2008.

Office of the Inspector General (2006). *A Review of the FBI's Handling of the Brandon Mayfield Case.* US Department of Justice, Washington, 330p. Available online at: http://www.usdoj.gov/oig/special/s0601/PDF_list.htm, accessed 12 January 2009.

Ogden N (2007). *VoicePay – so let's talk money.* Presentation at the Biometrics Exhibition and Conference 2007, 17–19 October 2007, Westminster, London.

O'Gorman L (1999). Fingerprint Verification. In A Jain, R Bolle and S Pankanti (eds.) *Biometrics: Personal Identification in Networked Society*, Kluwer Press, Dordrecht, p.43–64.

O'Hara K (2004). *Trust: From Socrates to Spin.* Icon Books Ltd., Cambridge, 256p.

O'Neill O (2002). *A Question of Trust: The BBC Reith Lectures 2002.* Cambridge University Press, Cambridge, 108p.

Organisation for Economic Cooperation and Development (1980). *Guidelines on the Protection of Privacy and Transborder Flows of Personal Data.* OECD, Paris. Available online at: http://www.oecd.org/document/18/0,3343,en_2649_34255_1815186_1_1_1_1,00.html, accessed 23 October 2008.

Organisation for Economic Co-operation and Development, Working Party on Information Security and Privacy (2004). *Biometric-Based Technologies*. Organisation for Economic Co-operation and Development, Paris, 66p.

Parliamentary Office of Science and Technology (POST) (2004). Radio Frequency Identification (RFID). *Postnote* 225: 1–4.

Parthasaradhi STV, Derakhshani R, Hornak LA and Schuckers SAC (2005). Time-Series Detection of Perspiration as a Liveness Test in Fingerprint Devices. *IEEE Transactions on Systems, Man, and Cybernetics – Part C: Applications and Reviews* 35(3): 335–343.

Perri 6, Lasky K and Fletcher A (1998). *The Future of Privacy. Volume 2, Public trust and the use of private information*. Demos, London, 144p.

Persaud KC, Lee D-H and Byun H-G (1999). Objective Odour Measurements. In A Jain, R Bolle and S Pankanti (eds.) *Biometrics: Personal Identification in Networked Society*, Kluwer Press, Dordrecht, p.251–272.

Phillips PJ, Scruggs WT, O'Toole AJ, Flynn PJ, Bowyer KW, Schott CL and Sharpe M (2007). *FRVT 2006 and ICE 2006 Large-Scale Results*. Technical Report NISTIR 7408, National Institute of Standards and Technology, Maryland, 55p.

Ponemon L (2006). *Global Study on the Public's Perception about Identity Management*. Ponemon Institute and Unisys Corporation, Michigan, US, 27p.

The Privacy Commissioner of Canada (2005). *Annual Report to Parliament 2004–2005*. The Privacy Commissioner of Canada, Ontario, 81p. Available online at: http://www.privcom.gc.ca/information/ar/200405/200405_pa_e.asp, accessed 17 November 2008.

The Privacy Commissioner of Canada (2006). *Government Accountability for Personal Information: Reforming the Privacy Act*. The Privacy Commissioner of Canada, Ontario. Available online at: http://wwwprivcom.gc.ca/information/pub/pa_reform_060605_e.asp, accessed 17 November 2008.

Privacy Rights Clearinghouse (2007). *How Many Identity Theft Victims Are There? What Is the Impact on Victims?* Available online at: http://www.privacyrights.org/ar/idtheftsurveys.htm#FTC, accessed 16 June 2009.

Privy Council Office Canada (2004). *Securing an Open Society: Canada's National Security Policy*. Privy Council Office Canada. Ontario. Available online at: http://www.pco-bcp.gc.ca/index.asp?doc=natsec-secnat/natsec-secnat_e.htm&lang=eng&page=information&sub=publications, accessed 17 November 2008.

Prokoski FJ and Riedel RB (1999). Infrared Identification of Faces and Body Parts. In A Jain, R Bolle and S Pankanti (eds.) *Biometrics: Personal Identification in Networked Society*, Kluwer Press, Dordrecht, p.191–212.

Pugliese J (2005). In Silico Race and the Heteronomy of Biometric Proxies: Biometrics in the Context of Civilian Life, Border Security and Counter-Terrorism Laws. *The Australian Feminist Law Journal* 23: 1–32.

Pulliam TJ and Schuster MM (1995). Congenital markers for chronic intestinal pseudoobstruction. *The American Journal of Gastroenterology* 90(6): 922–926.

Quinlan R (2008). More than 120 data-storage devices 'lost' by government staff. *The Sunday Independent* 27 April 2008. Available online at: http://www.independent.ie/national-news/more-than-120-datastorage-devices-lost-by-government-staff-1360451.html, accessed 27 April 2008.

Rachels J (1984). Why Privacy is Important. In FD Schoeman (ed.) Philosophical *Dimensions of Privacy. An Anthology*. Cambridge University Press, New York, p.290–299.

Rawls J (1993). *Political Liberalism*. Cambridge University Press, New York, 464p.

Rawls J (1994). *Ethics in the Public Domain*. Oxford Clarendon Press, Oxford, 250p.

Reiman JH (1984). Privacy, Intimacy, and Personhood. In FD Schoeman (ed.) *Philosophical Dimensions of Privacy. An Anthology*. Cambridge University Press, New York, p.300–316.

Riera A, Soria-Frisch A, Caparrini M, Grau C and Ruffini G (2008). Unobtrusive Biometric System Based on Electroencephalogram Analysis. *EURASIP Journal on Advances in Signal Processing* Volume 2008: Article ID 143728, 8p, doi:10.1155/2008/143728.

Robinson N, Graux H, Botterman M and Valeri L (2009). *Review of the European Data Protection Directive*, RAND Corporation, 82p.

Rommelaere J (2007). *Togo government accomplishes nationwide biometric registration of its voters*. Presentation at the Biometrics Exhibition and Conference 2007, 17–19 October 2007, Westminster, London.

Rowe RK (2005). A Multispectral Sensor for Fingerprint Spoof Detection. *Sensors* 22(1): 1–4.

RTÉ Business (2005). *Bank customers warned to be on their guard*. RTÉ Business published online 9 November 2005. Available online at: http://www.rte.ie/business/2005/1109/ipso.html, accessed 17 June 2009.

Rudin N, Inman K, Stolovitzky G and Rigoutsos I (1999). DNA Based Identification. In A Jain, R Bolle and S Pankanti (eds.) *Biometrics: Personal Identification in Networked Society*, Kluwer Press, Dordrecht, p.287–310.

Ryan FW (2002). *Constitutional Law*. Round Hall Ltd., Dublin 165p.

Sarkar S and Liu Z (2008). Gait Recognition. In AK Jain, P Flynn and AA Ross (eds.) *Handbook of Biometrics*, Springer, New York, p.109–129.

Sarkar S, Phillips PJ, Liu Z, Vega IR, Grother P and Bowyer KW (2005). The HumanID Gait Challenge Problem: Data Sets, Performance, and Analysis. *IEEE Transactions on Pattern Analysis and Machine Intelligence* 27(2): 162–177.

Seghetti LM and Viña SR (2004). *U.S. Visitor and Immigrant Status Indicator Technology Program (US-VISIT)*. CRS Report for Congress, Washington, 32p.

Sheikh AA (2008). *The Data Protection Acts 1988 and 2003: Some Implications for Public Health and Medical Research*, Health Research Board, Dublin, 130p.

Snijder M (2007). *Report on the Workshop Security & Privacy in Large Scale Biometric Systems*. European Biometrics Forum, Dublin, 28p.

Standing Committee on Access to Information, Privacy and Ethics, *Minutes of* Proceedings. Available online at: http://www2.parl.gc.ca/HousePublications/Publication.aspx?DocId=3426491&Language=E&mode=1&Parl=39&Ses=2, accessed 17 November 2008.

Stanley J and Steinhardt B (2002). *Drawing a Blank: The failure of facial recognition technology in Tampa, Florida*. American Civil Liberties Union, New York, 9p. Available online at: http://www.aclu.org/FilesPDFs/drawing_blank.pdf, accessed 20 February 2008.

Taylor C (2008). Thousands of social welfare details on stolen laptop. *The Irish Times* 11 August 2008. Available online at: http://www.irishtimes.com/newspaper/breaking/2008/0811/breaking25.htm, accessed 11 August 2008.

Teoh ABJ and Yuang CT (2007). Cancellable Biometrics Realization with Multispace Random Projections. *IEEE Transactions on Systems, Man, and Cybernetics – Part B: Cybernetics* 37(5): 1096–1106.

Thieme M (2008). *Business Potential for Genkey Technology*. International Biometric Group, New York. Available online at: http://genkeycorp.com/index.php?n=19&task=vis&id=5, accessed 15 February 2008.

Thomson JJ (1984). The Right to Privacy. In FD Schoeman (ed.) Philosophical *Dimensions of Privacy. An Anthology*. Cambridge University Press, New York, p.272–289.

Troitzsch H, Eschenburg F, Bente G, Krämer N, Lylykangas J, Vuorinen K and Surakka V (2005). *Deliverable D6.5. Introduction of a Multi-Modular Acceptance and Usability Questionnaire*. Version 1.2, 92p.

Uhl A and Wild P (2008). Footprint-based biometric verification. *Journal of Electronic Imaging* 17(1): 11–16.

United States Immigration and Naturalization Service (INS), Office of Inspections (nd). *INS Passenger Accelerated Service System (INSPASS) Briefing Paper*. Available online at: http://www.biometrics.org/REPORTS/INSPASS2.html, accessed 11 April 2008.

Van Camp N and Dierickx K (2008). The retention of forensic DNA samples: a socio-ethical evaluation of current practices in the EU. *Journal of Medical Ethics* 34(8): 606–610.

van der Ploeg I (1999). The illegal body: 'Eurodac' and the politics of biometric identification. *Ethics and Information Technology* 1(4): 295–302.

van der Ploeg I (2005a). *Biometric Identification Technologies: Ethical Implications of the Informatization of the Body*. BITE Policy Paper No.1. 18p. Available online at: http://www.biteproject.org/documents/policy_paper_1_july_version.pdf, accessed 16 October 2007.

van der Ploeg I (2005b). *The Politics of Biometric Identification. Normative aspects of automated social categorization*. BITE Policy Paper No.2. 16p. Available online at: http://www.biteproject.org/documents/politics_of_biometric_identity%20.pdf, accessed l8 November 2007.

Wayman JL (2000). Fundamentals of Biometric Authentication Technologies. In JL Wayman (ed.) *National Biometric Test Center Collected Works 1997–2000*. Version 1.2. San Jose State University, p.1–20.

Weber K (2006). *The Next Step: Privacy Invasion by Biometrics and ICT Implants*. Presentation given at the Zif Workshop on Privacy, February 2006. Available at: http://www.acm.org/ubiquity/views/pf/v7i45_weber.pdf, accessed 20 November 2008.

Weiser B (2004). Can Prints Lie? Yes, Man Finds to His Dismay. *The New York Times* 31 May 2004. Available online at: http://www.nytimes.com/2004/05/31/nyregion/31IDEN.html, accessed 23 February 2009.

Wickins J (2007). The ethics of biometrics: the risk of social exclusion from the widespread use of electronic identification. *Science and Engineering Ethics* 13(1): 45–54.

Wong R (2007). *Innovative use of biometrics in Hong Kong*. Presentation at the Biometrics Exhibition and Conference 2007, 17–19 October 2007, Westminster, London.

Woodward JD Jr (2001). *Super Bowl Surveillance: Facing Up to Biometrics*. RAND, California, 16p.

Woodward JD Jr, Webb KW, Newton EM, Bradley M, Rubenson D, Larson K, Lilly J, Smythe K, Houghton B, Pincus HA, Schachter JM, Steinberg P (2001). *Army Biometric Applications. Identifying and Addressing Sociocultural Concerns*. RAND, California, 185p.

Yan P and Bowyer KW (2007). Biometric Recognition Using 3D Ear Shape. *IEEE Transactions on Pattern Analysis and Machine Intelligence* 29(8): 1297–1308.

Yongo I (2007). Technology used to combat truancy. *The Irish Times* 25 April 2007.

Zhang D, Liu Z, Yan J and Shi P (2007). Tongue-Print: A Novel Biometrics Pattern. *Lecture Notes in Computer Science* 4642: 1174–1183.

Zunkel RL (1999). Hand Geometry Based Verification. In A Jain, R Bolle and S Pankanti (eds.) *Biometrics: Personal Identification in Networked Society*, Kluwer Press, Dordrecht, p.87–101.